Knowledge Games

Tech.edu:
A Hopkins Series on Education and Technology

KNOWLEDGE
GAMES

How Playing Games Can Solve Problems, Create Insight, and Make Change

KAREN SCHRIER

Johns Hopkins University Press
Baltimore

Johns Hopkins University Press
2715 North Charles Street
Baltimore, Maryland 21218-4363
www.press.jhu.edu

Library of Congress Cataloging-in-Publication Data

Names: Schrier, Karen (Karen L.)
Title: Knowledge games : how playing games can solve problems, create insight, and
make change / Karen Schrier.
Description: Baltimore : Johns Hopkins University Press, [2016] | Series: Tech.edu |
Includes bibliographical references and index.
Identifiers: LCCN 2015026984 | ISBN 978-1-4214-1920-6 (hardcover : alk. paper) |
ISBN 978-1-4214-1921-3 (electronic) | ISBN 1-4214-1920-3 (hardcover : alk. paper) |
ISBN 1-4214-1921-1 (electronic)
Subjects: LCSH: Games—Psychological aspects. | Games—Social aspects. |
Educational games.
Classification: LCC GV1201.37 .S37 2016 | DDC 793.01—dc23
LC record available at http://lccn.loc.gov/2015026984

A catalog record for this book is available from the British Library.

*Special discounts are available for bulk purchases of this book. For more
information, please contact Special Sales at 410-516-6936 or
specialsales@press.jhu.edu.*

Johns Hopkins University Press uses environmentally friendly book materials,
including recycled text paper that is composed of at least 30 percent post-consumer
waste, whenever possible.

To my family

CONTENTS

I want to thank the Marist College staff, faculty and administrators, including the Department of Media Arts and the School of Communication and the Arts. I am particularly grateful for the support of Dr. Sue Lawrence, Matthew Frieburghaus, Dr. Ron Coleman, Dr. Keith Strudler, Dr. Shannon Roper, Matthew Johnson, Eric Nuñez, Dr. Adam Zaretsky, Dr. Laura Linder, the SCA office staff (Cindy Miller, Robin Will, Laura Leahy and Monica Schott), Dr. Lyn Lepre, Dr. Thomas Wermuth, Dr. Geoffrey Brackett, and Dr. Dennis Murray. I also want to thank my current and former students for their enthusiasm for games.

I am appreciative of the insightfulness, passion, and camaraderie of the global games community, including the Learning and Education Games (LEG) Special Interest Group of the International Game Developers Association (IGDA) (including fellow steering committee members Dr. Stephen Jacobs, Matt Nolin, Dr. Elena Bertozzi, Dr. Brock Dubbels, and Dr. David Simkins), the New York City IGDA group, the Hudson Valley Game Developers Meetup Group (including Dr. Doug Maynard and Dr. Chris Garrett), and Game-Based Learning NYC (GBLNYC) (including Joe Ballou and Liz Jasko).

My ideas first took shape during my time as a graduate student at the Massachusetts Institute of Technology in the Comparative Media Studies program, and continued to develop while a doctoral student at Columbia University's Teachers College. I am grateful to the professors, mentors, and advisers who inspired me, including Dr. Henry Jenkins III, Dr. Chris Dede, and Dr. Charles Kinzer, as well as my fellow CMS and TC students.

Thank you to Greg Britton, Catherine Goldstead, Yasmine Kaminsky, Hilary S. Jacqmin, and Johns Hopkins University Press staff for their support, encouragement, and guidance in publishing this book, and to Glenn Perkins for copyediting the manuscript.

I want to thank my family, including my parents, Steven and Janet, and my brother, David, and my grandparents, Anne and Bernard, and Betty and Seymour. I also want to thank Bernard and Sandra Shaenfield.

I am in awe of the patience, love, and support provided by my husband, Dr. David Shaenfield.

This book is dedicated to my beloved children, Alyssa Leia, Noah Benjamin, and Jason Richard. During the writing and editing of this book, my son Noah Benjamin was born three and a half months early, after his twin, Jason Richard, died suddenly in utero (due to a number of unpreventable and as yet incurable complications related to identical twins sharing a placenta). Noah spent three grueling months in the NICU. Each day he bravely faced many obstacles, and each milestone—whether gaining an ounce, feeding by mouth, or breathing on his own—was hard won. This book was conceived alongside those heartbreaks and miracles, and it is likewise filled with the hopes and doubts that life brings. These events reminded me how much about our world is still unknown and unknowable. Will knowledge games help us solve life's greatest mysteries? Can they help us give meaning to its randomness? Will they give us insight into our dreams and our fears—or will some things always stay unknown? I look forward to seeking these answers.

Introduction

It was the summer of 1994 and a typical day in a biology lab at the State University of New York at Stony Brook. I pipetted a microliter of plasmid DNA into a small tube, swirled in some water, added a drop of buffer and a dash of enzyme, and let it digest for a while. Afterward, I poured the mixture into an agarose gel, watching it snake through the viscous jelly and separate into bands that could be seen under an ultraviolet light. Despite my daily routine of performing these restriction enzyme digestion and gel electrophoresis procedures,[1] I found it wondrous each time. Where would the DNA bands land today? What would I learn about the gene I was studying? What discoveries might I stumble on?

It was a normal day for a professional scientist—except that I was not a professional. At the time, I was only 16 years old. I had not even graduated from high school, let alone college, a doctoral program, or a postdoctoral fellowship—the typical path to being vetted and valued as a scientist. I was an amateur, a faker. Yet there I was, working in a real lab at one of the top science centers in the world. I was using authentic tools and materials, apprenticing alongside other scientists, and working on wholly unsolved problems related to insulin regulation that had actual implications for diabetes. I was contributing to science! Heck, I even wore a lab coat and goggles and needed an ID to enter the building.

If only there were more ways where kids and adults—of all backgrounds, perspectives, and levels of expertise—could have an opportunity to solve authentic problems, interact with real instruments, data, or systems, and provide valuable contributions to our societal knowledge.[2]

Has this day come? Perhaps.

Now anyone can log on to Galaxy Zoo, an online platform, and categorize images of galaxies, contributing to our knowledge of the universe. We can inspect birds in our backyard and enter our observations on eBird, along with thousands of other bird watchers. Or we can track and record cicadas in Cicada Tracker.[3] These activities are called "citizen science" projects because they enlist ordinary people in data collection, interpretation, or manipulation activities to collectively solve scientific quandaries. There are currently hundreds of public citizen science projects going on all over the world, many initiated in the past five years.[4] This trend does not seem to be showing any signs of slowing down.

Citizen science contributions do not have to be solely scientific in nature. More broadly, companies, institutions, and researchers have been increasingly recruiting participants from the public to help perform activities and donate time and effort in what is commonly called "crowdsourcing."[5] Crowdsourcing is a process of collecting contributions from the "crowd," where participants might be invited to do anything from provide opinions, collect and share digitized items, analyze data, or submit observations. For example, anyone can annotate and tag war diaries from World War I at Operation War Diary. While visiting What's the Score at Bodleian?, the public can thumb through Oxford University's digitized Bodleian musical score collection and provide metadata on everything from a cover image to the types of instruments used in the score. Or people can use a mobile device to submit reports of their sexual activity and general location to the Kinsey Reporter app—all for important research purposes.[6]

What's more, people are now playing video games—of all things—to participate as citizen scientists and crowdsourcers and through these games helping to solve real-world problems that scientists and other researchers have posed. Game players can work on puzzles and fold 3-D proteins to better understand the structure of real proteins in *Foldit*. People can analyze cancer data in a colorful mobile puzzle game in *Reverse the Odds*. In the Defense Advanced Research Projects Agency's *Monster Proof*, players can interact with a virtual monster village, solve logic puzzles, gather and trade resources, and build roads while simultaneously checking software source code to ensure that it will work properly. And Parrot AR.Drone Quadricopter owners can use the drone, coupled with an iOS app, to play an augmented reality game, the *Astro Drone Crowdsourcing Game*, where they perform space missions to teach scientists better ways to manage and dock space probes in real life.[7]

Through games, people could identify biases, observe and record emotions, analyze specimens, solve puzzles, make estimations, describe social interactions, make ethical decisions, probe simulations, launch hypotheses, experience consequences, and provide points of view on issues and policies. Game players could be collectors, contributors, purveyors, and interpreters of scientific, humanistic, social, and psychological data and evidence. In other words, through these types of games, anyone, not just vetted scientists or researchers, can make valuable contributions to our common knowledge and perhaps even help solve some of life's mysteries.

In this book I unravel the potentials, pitfalls, limits, questions, and implications of solving problems and producing new knowledge through games. Humanity faces very pressing, complex problems. For example, although we have the know-how, scientifically and technologically, to feed, heal, educate, clothe, and shelter the people of the world, we have not been able to do so because of social, cultural, or geographic constraints, to name but a few.[8] Communicable diseases require not only scientists and doctors to analyze them but also epidemiologists and social scientists in partnership with the public to understand how these diseases spread, which social mechanics they exploit, and which types of people are more susceptible. Societal problems and global concerns, such as depression, bullying, posttraumatic stress disorder (PTSD), conflict and genocide, require both broad interdisciplinary thinking and deep attention to physiological, sociological, humanistic, and psychological factors. Moreover, real-world change and innovation is becoming more and more difficult to achieve, as we require increasingly specialized knowledge, coupled with interconnections with other fields.[9] Imagine if new insights came not just from research centers, companies, think tanks, and universities but from games as well. Could games help us solve complex interdisciplinary problems, produce new knowledge, and make effective change?

The Games Are A-Changing

The past few decades have seen rapid growth in gaming and the game industry, as well as an expansion in the role games play in our everyday lives. According to the NPD Group and the Entertainment Software Association, there were 273 million units of computer and video games sold in 2009, amounting to more than $10.5 billion in revenue for the industry. In 2014 a large proportion of time in the United States was spent playing video games, and around 59% of Americans played them. The average game player of 2010 spent about 8 hours a week playing video games. In 2011 a popularly cited statistic was that by the time a typical kid is 21 years old, he or she has spent over 10,000 hours playing video games, which Jane McGonigal notes is "24 hours less than [kids] spend in a classroom for all of middle and high school if they have perfect attendance."[10] As video games become more popular, public concern over their use seems to be growing. For instance, a current zeitgeist is that games (and gamers) are not only *not* solving anything but are actively *causing* some of society's greatest ills. Games are sometimes seen as, at best, a waste of time and are often perceived and described as hotbeds of violence, addiction, sexism, racism, homophobia, and laziness. A recent *Law and Order: Special Victims Unit* episode dramatizes some of the current game controversies, such as Gamergate, doxxing, and online harassment. In his book *Everything Bad Is Good for You*, Steven Johnson observes that "you can't get much more conventional than the conventional wisdom that kids today would be better off spending more time reading books, and less time zoning out in front of their videogames."[11]

At the same time, we have also been increasingly deliberating the efficacy of using games for learning and education, for social change, and other so-called serious purposes. For example, we now have games that purportedly support learning everything from civics to physics, from empathy skills to math facts. Games are being used for military training, for teaching surgical procedures, to raise awareness of global issues such as domestic violence and oppression, and for practicing how to talk to veterans suffering from PTSD. Leading education researchers such as James Paul Gee, Kurt Squire, Katie Salen, Eric Klopfer, Mizuko Ito, and Constance Steinkuehler have explored the literacy, social scientific, and scientific inquiry practices related to games, including mainstream games and those more explicitly designed for educational purposes. In his excellent book *How Computer Games Help Children Learn*, David Shaffer emphasizes how in-game learning, practice, and problem solving can help people generate the skills and knowledge needed to make innovative changes in the world.[12]

It is time to further these conversations. Rather than just helping us *learn* that which is already known, can games also *reveal* the unknown? Rather than showing us *how* to make real-world change, can games actually *be* the change we want to make? Rather than *causing* problems, can games help *solve* them? Rather than helping us destroy, can games help us build? Rather than helping us exclude, can games help us embrace? Imagine if some portion of the 3 billion hours a week we spend, globally, playing games could help us create new knowledge, solve problems, make change, or even love, empathize, wonder, dream, or inspire?[13]

- Could we couple games with real-world problem solving and invite players to interact, observe, feel, express, inspect, identify, empathize, analyze, judge, and behave from within a scalable, authentic, and living, breathing system?

- Could we elicit collective intelligence and distributed cognition by encouraging people with varying expertise and perspectives to play together and contribute skills and abilities, enabling a jumble of play that produces more than the sum of its parts?[14]
- Could games provide possible paths toward solving a problem, and even help find solutions?
- Could game players produce new insights in the scientific, social scientific, artistic, and humanistic fields—on everything from cancer and AIDS to historical trends and color perceptions, bullying and ballads, ethics and entomology, dialects and dog behavior?
- Could the playing of a game actually help alter our workplaces or schools, communities, and institutions and make widespread changes?
- Could these games even help players to change themselves by encouraging them be more empathetic, creative, or inventive, or even less likely to drop out, overeat, or drink and drive?

If so, what are the complexities and implications of designing and using games for knowledge production and problem solving? What are the questions we need to ask before we create and deploy these games? For example, why might these games be effective? What are the unseen consequences of their use? What are the appropriate outcomes of these games? Could we (and should we) use the knowledge produced by a game to motivate people to vote, to know when and how to help friends and family suffering from mental illness, to find new ways to reduce carbon emissions, or to increase retention in college? Should we redesign a game experience—in real time—for a particular individual or a larger community to help motivate real-world changes? How might these practices redefine games or even change the role of games in our society? Conversely, do these types of games reinvent what it means to discover and to *know*? Will they transform how we create, share, and establish new knowledge as a society?

This book considers the psychological, historical, logistical, design, social, cultural, ethical, and economic dimensions and implications of these types of games. What are the critical questions and challenges that loom as we toddle into a future of using games to produce knowledge and solve problems?

Currently, the games I describe—ones that potentially enable people to contribute, solve authentic problems, produce new ways of seeing the world, and make real-world change—do not have a consistent or cohesive name.

These games that focus on scientific problems have been called "citizen science games" or "crowd science games," while others have more broadly used the terms "games with a purpose" (GWAP), "social participation games," "crowdsourcing games," "applied problem solving games," and "human computation games" because they have some greater purpose or goal, solve problems, and they use the collective participation of human players to support computational activities.[15] Yet none of these terms seems general or precise enough to encapsulate and properly address these types of games. For the purposes of brevity and clarity in this book, I use the term "knowledge games" to label and express those games that seek to invent, create, and synthesize new understandings of the world, solve real-world problems big and small, and help us reconsider, reframe, and reflect on humanity and our universe. These games solve problems and create new understandings inside the game, but with real-world application. Although to some extent all games could provide novel glimpses into humanity, only those games with a primary goal of producing new knowledge fit the term "knowledge games" as I use it.[16] I use the term "games," rather than "video games" or "digital games," as I believe nondigital or digital games could also be considered knowledge games. In this book I typically discuss examples of digital games, but I invite others to design, consider, use, and evaluate nondigital knowledge games as well.[17]

What This Is and Isn't

The focus of this book is not on games in education, such as how play should be conducted in schools or other learning environments. Nor is it on curricular considerations, metrics for academic success, or design principles for learning games. I am not arguing that people should design or play more so-called serious games or games for learning—whether classic educational games such as *Oregon Trail* or *Math Blaster* or newer games such as *Mission US* or *Quandary*—or that they should modify commercial-off-the-shelf games such as *Angry Birds*, *Portal*, *Sim City*, *Minecraft*, or *Civilization* for educational purposes.[18] Rather than consider how kids can learn more about astronomy or adults can hone interpersonal skills through games (such as in SciGames.org's *Planet Families* or Kognito's *Family of Heroes*), I focus on how we can use games to create knowledge about our universe or develop original insights into human interactions. That said, in this book, I consider current research on games and education to highlight ways games could be used for *both* learning and knowledge production, problem solving and real-world application, as they often go hand in hand.

This book is also not centered only on "games for change" or "social impact games" or games that typically present some social, economic, political, historic, education, or scientific issue—such as the challenge of achieving peace in the Middle East—and are created to help spur knowledge, understanding, or even real-world action related to this issue.[19] Examples of games for change include *Peacemaker*, *Darfur Is Dying*, *The World Peace Game*, *Ayiti: The Cost of Life*, and *Half the Sky*. An important distinction is that with games for change, the player may learn about an issue, and even understand how to take action or learn how to solve a problem related to it, but the action or problem solving can take place *outside* of the game. The playing of the game reveals an argument about an issue, but it does not necessarily construct a possible solution or new perspective on the issue. The games I describe in this book are not just *for* change, they *are* change. Knowledge games are not just contributing to an individual's knowledge of an issue, but they also aid us in overhauling, reapproaching, or reconceptualizing the issue itself. While knowledge games could be considered a subtype of a game for change or social impact game, the reverse is not always true.[20]

I am also not trying to legitimize games by making them more edifying, more meaningful, or more socially valuable. In other words, my purpose is not to argue that games, gamers, game players, the game industry, or game culture are (or should be) constructive, useful, and innovative, and able to aid learning, social change, knowledge production, or problem solving. Nor could we claim that games are universally good or bad for society simply because a few of them, or even hundreds of them, may help solve life's mysteries and puzzles. Games have many aims, meanings, and uses, just like any medium, tool, or experience. There are a variety of ways to use games, such as for pranks, exercise, art, or work, which could range from nefarious and criminal to edifying and inspiring, entirely entertaining to tediously laborious.[21] People do not need to be hamster gamers, constantly generating electric ideas for society—no one needs to be a play martyr. Nor do we need to only design games that solve complex problems, contribute to society, or try to do "good" for the world. Numerous designs, motivations, desires, and play possibilities are acceptable, and one is not necessarily more appropriate than another. Games and game players *do not* need to solve problems to make them matter—they already matter—but if they *can* also help us solve real-world problems, we should understand their opportunities and obstacles.

Although games are a prominent part of this book, I am not calling for *everything* to be more gamelike. This book is not about gamification or gamifying knowledge production. I am not asking for all scientific, humanistic, and social scientific research or all problem solving, knowledge creation, government, and democratic processes to be gamified so that we are living life in a game and that we are all game players, all the time. Regardless of whether understanding ourselves more deeply would be more "fun" in games or a "gamified" context, not all problems are best solved or all insights are best gained through games. Knowledge games may be another viable way we produce new knowledge, but they are not necessarily a better way. As such, it is important to understand their unique advantages and disadvantages as compared to other knowledge-seeking experiences and methods. Knowledge games, as I have defined them, are not a form of gamification. Gamification is popularly considered the process of taking game elements or gamelike features (such as points, badges, levels, and missions) and applying them to nongame contexts, such as office tasks, classroom activities, or hospital interventions, with the goal of, for example, increasing productivity in the workplace, enhancing the learning of a novel, or improving compliance with medication. These initiatives may or may not be effective, but assessing whether gamification techniques are effective is not the purpose of this book.[22] Gamification involves adding game elements to known real-world problems, whereas knowledge games involve the full integration of that problem into a game. In knowledge games, the problem *is* the game. *EyeWire* does not take game mechanics and apply them to studying neurons or neurological substrates, nor does it take these elements and tack them onto science research. Rather, in *EyeWire*, neurons and the process of investigating them are transformed into a game environment, and we can use this to directly produce new knowledge with real-world applications. Thus, knowledge games are *games* first and foremost, which means that a key challenge for knowledge games is providing an appropriate player experience alongside the pursuit of new insights and solutions. This book considers how to design effective, engaging games that also generate novel solutions to real-world problems.[23]

Finally, to fully explore how games are or are not reinventing knowledge production or problem solving, we need to be open to all game-design methods, genres, types, and mechanics. Applying hierarchies of taste to games can only detract from our understanding, as any type of game, genre, or mechanic may be appropriate for eliciting collective problem-solving action and all game types should be considered and assessed. The game mechanic, genre, or goal used is not the message and should not prescribe its purpose or its implications. As far as we know, a social alternate reality game (ARG) could help people work together to solve a scientific mystery just as readily as a puzzle platformer or a match-3 social mobile game could help us study cancer or realign our healthcare system—it all depends on the audience, context, problem, project goals, and type of play experience for which we are designing.[24]

Why Now?

So why should we discuss knowledge games now? As of early 2016, there are less than a hundred, and perhaps even fewer than a couple dozen, games that could be considered members of this category. The knowledge games that exist—such as *Foldit, EteRNA, Reverse the Odds, Monster Proof,* or *Apetopia*—are not necessarily good, well-designed, engaging, or even effective knowledge-making games, nor are they all successful or useful (though many of them are effective and engaging).[25] Moreover, they are a mere drop in the bucket in terms of what is happening today in the game industry, in game development, or in the well-rooted realm of research and knowledge creation.

But this is changing. While the conversation this book proposes may not seem urgent right now, it is necessary. Although few real-world enigmas are currently being solved through games, the very existence of knowledge games stirs up new—and silently simmering—critical questions to consider as a society. Moreover, in the future, knowledge gaming may very well be a significant way that we innovate and contribute to our knowledge of humanity and the universe where we live.

To start this important conversation, throughout this book, I consider how knowledge games can enable and encourage problem-solving processes, and what types of interactions, ecologies, and experiences may be useful for this to occur. Which design principles support their creation and use?

I also pose the guiding questions that we need to consider as we increasingly design and implement knowledge games. For example, how should we couch the conclusions we draw from the large data sets generated from people doing authentic tasks and activities in a system like a game? Can we motivate people or achieve equitable participation in games so that we get a wider range of diverse views and perspectives? What are the implications of people spending their leisure time contributing time and effort, and what are the boundaries (if any) between work and play? When do knowledge games become inappropriate or exploitative uses of players?

As more and more people are becoming media producers, problem solvers, game players, and contributors, how might the role of the public in knowledge making evolve? If games are a way to solve social and scientific problems, contribute knowledge, and make change in our communities or our institutions, does this redefine democracy? Is participating in these games a new way to engage civically? Are knowledge games supporting democratic participation, or are we just replicating hegemonic relationships that already exist, such as between scientist and amateur, or designer and player? To fully participate in society, must we game?

Although I ask these critical questions throughout this book, I also recognize that there are no clear answers. I hope this book advances the search for them.

Mapping the Book

Just as we need interdisciplinary, iterative, innovative, and participatory approaches to solve complex real-world problems, we also need such approaches to critique and address knowledge games. In this book, I consider research, frameworks, and theories from diverse fields, including games, citizen science, and crowdsourcing, as well as a range of disciplines, such as psychology, education, critical theory, computer science, design, human-computer interaction, media studies, history, and economics.[26] I realize that not everyone will be satisfied with such an interdisciplinary pastiche, but using multiple lenses is necessary for such a multifaceted phenomenon. That said, not all perspectives are covered, so I look to you to fill in those gaps.

In each chapter, I dive deeply into one theme or research area related to knowledge games, such as motivation, participation, design, and data. To illustrate these topics, I explore at least one relevant knowledge-producing game or platform in detail in each chapter. Although I am critical in my analysis of the games mentioned in this book, I respect this nascent field and realize the immense challenge of designing these new types of experiences. Moreover, because of the newness of this field, there is little research related to knowledge games or practical guidelines on how to design or use them. All findings and evaluations should therefore be considered preliminary, and I invite you to appraise and add to them, and to contribute new perspectives and evidence. To help further the conversation around knowledge games, I pose and grapple with many questions, which are summarized in the appendix. In the appendix, I also provide an outline of possible design principles that could be used for creating, revising, and implementing knowledge games, and which should be furthered studied and empirically tested.

In part I, I introduce the major terms, concepts, processes, and research realms related to knowledge games.

In chapter 1, I introduce the concepts of crowdsourcing and citizen science, processes related to knowledge games. How is the public already contributing to research and to knowledge production in general? How are institutions and companies incorporating public perspectives and opinions, as well as problem-solving abilities, into their businesses and research agendas?

I bring games into the investigation in chapter 2. What are the key aspects of games, and the way they are designed, which might herald their use as knowledge-creating experiences? What is design and how is a design perspective useful in creating, using, evaluating, and interpreting knowledge games?

In part II, I consider three areas of research on games—problem solving, motivation, and social interaction—to understand how to better design and deploy games for knowledge production and problem solving.[27]

In chapter 3, I consider educational, psychological, and cognitive science research on problem solving and how games might be particularly relevant (or limited) in supporting the processes around problem solving. How can we create an effective game environment for problem solving?

I discuss the research surrounding motivation in chapter 4, including what motivates volunteers, crowdsourcers, and citizen scientists, and use this in conjunction with motivation research on games. How can we motivate enough (and the right people) to play knowledge games?

In chapter 5, I cover the different types of social interactions within and surrounding games, including cooperation and collaboration. What are the ways that games might be able to support (or limit) social interaction, such that we can better solve problems and create new knowledge?

In part III, I investigate issues related to knowledge gaming, including relationships between amateurs and scientists, leisure and work, and producers and consumers; changing perceptions of research, analysis, and data; and how the role of the public in knowledge production may (or may not) be changing.

In chapter 6, I explore the history of amateur participation in citizen science and in scientific problem solving and collection activities, including shifting perceptions in what it means to be an expert and who is an amateur.

I consider whether the relationship between scientist and amateur is truly changing and how the past can help us understand knowledge production today and in the future.

I consider conceptions of participation in media and knowledge production in chapter 7. Who is participating, and what are ways that people can or cannot access and participate in knowledge gaming? How can we characterize participation in these games—are they leisure or work, and when is the participation appropriate, or even exploitative?

In chapter 8, I delve into changing practices and perceptions around research and data analytics. I discuss trends in data collection, interpretation, and manipulation, through games and other media, and how it is being used to understand and predict human behavior for both useful and nefarious purposes, and everything in between. Could this kind of data be used to gain insight into human nature—and what are its limits?

Finally, in chapter 9, I consider the concept of knowledge. Is the process of knowledge production changing, and, if so, what is the potential role of games in this transformation? How might knowledge games redefine what games are—or even what it means to be engaged citizens of the world? Will we need to "game" to be good citizens?

PART ONE

WHAT ARE KNOWLEDGE GAMES?

CHAPTER ONE

Contribution

Can we crowdsource sex? Alfred Kinsey's 1940s survey of sexual behavior was the first vigorous study of human sexuality. About 75 years later, the Kinsey Institute released Kinsey Reporter, a mobile app and online site where women and men from all over the world can contribute data on their sexual encounters. On the app and website, in real time, we can voyeuristically look at a map of these tagged rendezvous. For example, on Valentine's Day, February 14, 2015, there were reports such as "couple, man, woman, young, lust, happiness, satisfaction" tagged in Westbrook, Maine, United States, and "solo, man, young, love, lust, happiness, fear" recorded in Edinburgh, Scotland. These openly available data sets are then aggregated and can be downloaded and visualized on the site. Users can view reports on everything from flirting to fetishes, and can take periodic surveys. For instance, in one past survey conducted through the app, participants were asked which types of sexual activities they most desire on February 14. Anyone can view the report of the results—called "Valentine's Day"—and learn that 166 of the respondents want kissing and 164 want cuddling.[1]

While it may seem like intimate details are wildly being shared, the creators of the Kinsey Reporter app couch these contributions as coming from citizen scientists rather than exhibitionists. Instead of looking at the birds or the bees (such as in eBird or ZomBee Watch), contributors are invited to inspect themselves, their relationships, and their sexuality—all for the good of science. The Kinsey Reporter research team hopes to, for example, understand the incidence of unwanted sexual violence throughout the world or the relationships among birth control use and various sexual practices or locations.[2]

But how many people are really contributing to the Kinsey Reporter? Are the participants representative of the broader population? Are these contributions truly helping us understand more about human sexual behavior?

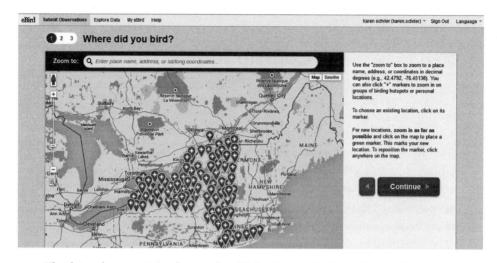

The data submission interface on the eBird website; eBird is a citizen science project where participants provide data about bird species they have observed in their location. Source: http://ebird.org/ebird/submit/map

Querying the Crowd

In general, people are spending lots of time contributing online by posting personal photos, commenting or tweeting, and liking, upvoting, or even providing opinions, money, or efforts through crowdsourcing, crowdfunding, or citizen science platforms. In this chapter, I introduce and detail some of the terms and concepts related to communal contribution—particularly those related to knowledge production—such as crowdsourcing, citizen science, crowd science, and human computation. I also apply these terms to games in an effort to holistically label the process and practice of creating new knowledge through games.

Crowdsourcing first became a neologism in Jeff Howe's "The Rise of Crowdsourcing," a 2006 *Wired* article. In his 2008 follow-up book, *Crowdsourcing: Why the Power of the Crowd Is Driving the Future of Business*, he defines crowdsourcing as "the process by which the power of the many can be leveraged to accomplish feats that were once the province of a specialized few" and equates it with democratic and meritocratic processes.[3] Since then, the definition and usage has morphed, and crowdsourcing has been popularly applied, sometimes inappropriately, to different types of collective activities supported and accelerated by networked technologies—including microtasks on Amazon's Mechanical Turk platform, designing and voting on T-shirt designs on Threadless, answering questions on Quora, or co-editing pages on *Wikipedia*. In his 2013 book *Crowdsourcing*, Daren Brabham attempts to narrowly define the term, citing crowdsourcing as an, "online, distributed problem-solving and production model that leverages the collective intelligence of online communities to serve specific organizational goals." In other words, organizations are using crowdsourcing to invite people to do the tasks and activities, or provide the opinions and perspectives, which they cannot accomplish alone. The online communities are the "crowd" part of the crowdsourcing concept, and they are enabled by an organization to participate in these activities, which can range in size, shape, type, needs, impact, and goals. Brabham argues that with crowdsourcing, power must be shared between the corporate stakeholders or organizers and the problem solvers, creators, writers, or testers. There needs to be benefits for each side, so that each is more or less equally benefiting. Thus, for crowdsourcing to truly be crowdsourcing, Brabham believes it needs to be a shared process between organizer and crowd.[4]

Brabham identifies two platforms, Threadless and InnoCentive, as examples of crowdsourcing. On Threadless, users can submit T-shirt and sweatshirt designs, and they can also vote on which ones should be printed and sold through the site. InnoCentive is a problem-solving platform where various science organizations, such as laboratories and research centers, post open problems, along with the requirements and any funding that would be provided upon a successful solution. InnoCentive participants can work together or independently to solve problems and have effectively solved over 1,500 problems of the 2,000 posted.[5]

Other examples of crowdsourcing-like activity have also been cited, some even before the term was widely used. Furthermore, people are not necessarily organized by a company or through a platform specifically designed for crowdsourcing. Sometimes a crowd self-organizes or works in tandem with more official efforts. Henry Jenkins discusses the role of what he calls the "collective intelligence" of fans working together to spoil the reality show *Survivor* and, essentially, crowdsourcing clues to determine, before the result was actually broadcast, which contestant had won the show's reward. Another example of crowdsourcing occurred after the 2014 disappearance of Malaysia Airlines flight MH370 in the southern Indian Ocean. In addition to the official search efforts, people from all over the world used Tomnod, a platform for viewing DigitalGlobe's satellite images of Earth, to look at available satellite images of the Indian Ocean in the hopes that someone would find the missing plane or other clues as to its whereabouts. Tomnod was also used after the 7.8 magnitude earthquake in Nepal to help find missing people and determine which locations needed supplies. Anyone could compare old and new satellite images and label locations of damaged roads, buildings, homes, and areas. As of April 28, 2015, more than 16,500 volunteers had placed 74,000 tags on satellite images. Similarly, people and organizations are contributing data and analysis time using web environments such as the Humanitarian Data Exchange (HDX), which are being used to support relief efforts in Nepal.[6]

Sometimes people contribute for more personal reasons, with collective results. The Ovia website and mobile app by Ovuline motivates people to track their fertility-related data.[7] You can input your menstrual cycle, body temperature, and the character of your cervical mucous, and the app will provide you with the most likely times you can get pregnant, as well as charts of your data over time and reminders of when to have sex. The app also collects these data anonymously to aggregate them and use them to better understand human fertility more generally, and to also make more accurate predictions about fertility for individual participants.

But what about popularly cited examples of crowdsourcing such as *Wikipedia*? Brabham argues that certain activities that have been deemed "crowdsourcing" are in fact not crowdsourcing. He contends, for example, that *Wikipedia* is not crowdsourcing because there is an imbalance between the organization's (top-down) and crowd's (bottom-up) power (in this case the power is too heavily on the bottom), whereas examples such as Threadless are more balanced between the organizers and the crowd and better reflect the definition of crowdsourcing by Brabham. Other potential crowdsourcing examples—such as Frito-Lay's annual "Do Us a Flavor" contest, where participants submit ideas for a new flavor of potato chips (such as last year's winning Wasabi Ginger)—seem to be too heavily weighted at the top and do not fit Brabham's definition.[8] In this instance, Frito-Lay curries approximately 14 million ideas from the public, makes most of the decisions about who wins, markets its Lay's brand, and, we assume, earns profits from the increase in number of chip bags sold, including those special "limited-time" chip flavors.

On the other hand, an example that better fits Brabham's crowdsourcing definition might be Google's Waze, a traffic and navigation app drivers and passengers can use to contribute to reports of, and view tags by, fellow drivers on accidents, roadside hazards, construction sites, and navigation changes on the road. Recently, when my husband was stuck in traffic on I-84, I was able to go to the Waze app, look up his approximate location, and find out that there was a car fire just ahead causing the standstill. With Waze, Google can also learn about road changes and use this to update its maps, and drivers can learn about accidents and use this information to adjust their routes.[9] Thus, both Google and the public are ostensibly benefiting from the app.

Finding examples where the relationship between participants and organizers is truly symbiotic, however, is problematic theoretically and pragmatically challenging. Even with the Waze app, stakeholders and owners can get conflated with users and participants, altering the balance of power. For example, if we consider the relationship between the drivers, and a tangential stakeholder, police officers, a new dynamic emerges. A recently added feature on Waze involves the ability to tag speed traps and officer locations, which gives additional power to drivers, upending the police-driver dynamic. At the same time, police officers are subverting this power by tagging fake speed traps and police sightings.[10]

While Brabham's focused definition of crowdsourcing lets him put a stake in the ground to define the boundaries of crowdsourcing, it relinquishes him from having to problematize the relationships between crowdsourcing organizers and participants. In other words, if the so-called mutual benefit between the two groups is not perfectly equal, he can simply designate it as not a crowdsourcing platform, instead of building an analytical framework to reflect on the complex and complicated relationships that might exist in crowdsourcing processes. Brabham's crowdsourcing definition also becomes trickier as we apply it to games. If we use as our definition that both the crowd and the designers, or users and owners, must be truly and equitably balanced in terms of power, access, transparency, or benefits, there may be no games left to meet the criteria. That said, Brabham's definition is a useful starting point.

Despite Brabham's useful attempt at a comprehensive yet limiting definition of crowdsourcing, other researchers (and the public) use the term in multiple, sometimes contradictory, ways. For example, two key crowdsourcing researchers, Zhao and Zhu, cite *Wikipedia* as a classic example of crowdsourcing. Zhao and Zhu conducted an analysis of research articles from 2006 to 2011 that use the term, and they found no consensus in how it is used or defined. The 55 different articles they looked at use "crowdsourcing" to refer to a process, paradigm, or a platform and focus variably on its systemic, conceptual, or contextual aspects. Zhao and Zhu also noted that the term "crowdsourcing" itself relies on the combination of two terms—"crowd" and "outsourcing"—that themselves are ambiguous and may not relate to what the term together describes.[11]

Citizens, Unite

Perhaps the term "citizen science" is more in line with the activities, processes, and relationships described in this book. "Citizen science" is related to crowdsourcing but more explicitly invites the participation of so-called amateurs in scientific mystery solving and the production of science, technology, engineering, and mathematics (STEM) knowledge. Prestopnik and Crowston call citizen science a "form of social computation where members of the public are recruited to contribute to scientific investigations" and where citizens and scientists become temporary collaborators. Often, citizen science involves many people contributing observations, classifications, or interpretations using mobile or web-enabled technologies, such as a website or app; however, the entire process could be nondigital. Citizen scientists work on real-world science-related conundrums by participating in portions of a research project, such as defining and deciding on the project's goals and questions, the data collection or classification activities, or the analysis and interpretation of the information generated.[12]

The everyday citizens of the world have classified bugs, cleaned up lakes, and even observed gravestones.[13] Many citizen science projects have been created to motivate the completion of the activities, puzzles, or tasks that a large number of human beings would need to perform—ones that a small group of scientists could not logistically, financially, or practically undertake themselves. This amateur participation, through a citizen science platform, can communally help to solve scientific issues or build new STEM knowledge, which would take otherwise many resources, such as people, computer power, money, or time, to complete. As a result of these projects, Mueller, Tippens, and Bryan describe that a relationship emerges between the public and scientists, who together become engaged as partners in research, which hopefully leads to both shared knowledge and an active, engaged community.[14]

Project FeederWatch, initiated at Cornell's Ornithology Laboratory, is one of the earliest models for Internet-enabled crowdsourcing or citizen science. The project invites people to check their backyards, local nature areas, parks, or community areas in North America to identify, survey, and count the birds they see during the winter. This helps the researchers track birds in different locales and see broader migratory movements and distributions, as well as longer-term population declines or growths.[15]

There are many other examples. In a citizen science project called Marine Metre Squared (Mm^2), participants in New Zealand are asked to monitor beaches and the marine environment to identify and record plants and animals. In National Geographic/JASON Project's SharkFinder, citizens get kits with a bunch of sediment and can sift through it to find fossils of sea life, such as shark fossils. The citizens then return the fossils for analysis by paleontologists. In Nerds for Nature's Mt. Diablo Morgan Fire project, participants need to take images of the summit of Mt. Diablo, a 3,000-acre mountain in California that burned in a fire in 2013.[16] The purpose of the project is to help scientists monitor the changing landscape and ecologies of this large location after the fire. In LibCrowds, people can help match, transcribe, tag, and translate cards from the British Library's card catalog. This is just a taste of the types of projects currently available. Over 100 citizen science projects are listed on SciStarter.org, for example. Citizen science is becoming prevalent enough that it now has conferences dedicated to its study. The inaugural Citizen Science Conference took place in 2015, organized by the Citizen Science Association.[17]

What's in a Name?

Besides "crowdsourcing" and "citizen science," there are other terms that attempt to describe similar phenomena, such as "collective creativity," "networked science," "democratized science," and "human computation." The term "human computation" was first used in 1838 and then applied by Luis von Ahn in his 2005 thesis to explain activities that use "human processing power to solve problems that computers cannot yet solve."[18] The term "human computation" refers to human beings and computers working in tandem to more efficiently solve problems or generate new knowledge, by each doing what they do better. In other words, human computation experiences invite people to participate in and conduct those activities where human beings specifically excel, such as with image or visual recognition (as in the case of the *ESP Game*/Google Image Labeler or reCAPTCHA), language understanding and transcription (as in the case of Soylent), or manipulation of 2-D and 3-D models (*Foldit* and *EyeWire*), in partnership with a computer, which performs the tasks that computers do particularly well, such as the speedy processing, filtering, and comparison of data.[19]

With so many terms to describe different flavors of the same processes, it's hard to choose one name to use when discussing what happens when people contribute collectively, particularly as these terms and the phenomena they describe evolve and adapt over time.

As I posed in the beginning of the chapter, what happens when we apply these processes to games? What should we call the types of games that produce new knowledge and rely on players to contribute? Some researchers have used the terms "crowdsourced games" and "crowdsourcing games." These terms were used to inspire the 2014 "Games & Crowds: Using Computer Games to Maximize Crowdsourcing Outcomes" conference in Washington, DC, one of the first official gatherings to specifically explore these types of games.[20] Appropriating crowdsourcing in this way, however, may not be precise enough to describe those games that produce knowledge with the help of many participants, particularly as the trendy term "crowdsourcing" is often misused and misapplied.[21] Also, while some researchers are using "crowdsourcing" to describe games that are enabling participants to contribute tasks and analyses for larger real-world goals, a quick Google search reveals that many people use this term to explain video games that were crowd*funded*, or collectively funded by multiple people. Any genre or type of game could be crowdfunded, making it confusing whether the term refers to a game that was financially supported by the public or is contributing to the public through play.

At the same time, there are many benefits to applying the term "crowdsourcing" to these types of games. It implies that these games invite a crowd and suggests that multiple people can participate and contribute to them. Moreover, the term is well known and generally not intimidating to the general public.

However, while the phrase "crowdsourcing games" benefits from the current popularity of the term "crowdsourcing," its ambiguity and imprecision, not to mention vast differences in how it is being applied and what it constitutes (and, if used narrowly à la Brabham's definition, then no games may qualify), make it a less than desirable term for the games described in this book.

Despite any shortcomings of using the term "crowdsourcing" to identify these types of games, the research on crowdsourcing—including findings on the motivation of people involved in crowdsourcing, typologies of crowdsourcing, design principles for creating crowdsourcing experiences, and the accuracy and quality of crowdsourced activity—all provide a useful context for understanding these types of games.

What other names could we consider instead? Researchers have used the terms "citizen science games," "crowd science games," or "human computation games." These terms are perhaps are too constrictive, as they imply a focus on STEM knowledge production or suggest that a game's activities are technological or technical, or that they require special expertise or equipment.[22] For example, while the term "human computation games" seems inclusive of all human beings, does not advance any topic or use (such as scientific knowledge), and suggests a symbiotic relationship between human and machine, the term "computation" is unwieldy and potentially discouraging to a broad audience. The terms "citizen science games" or "human computation games" may also inadvertently bias against the use of games for more humanistic or social purposes or may disproportionately draw in more people specifically interested in computers or computation. Moreover, "human computation" also implies that all of these types of games must marry human beings with computers for an optimal problem-solving relationship. This excludes nondigital games or even digital games that use other types of players (nonhumans, drones, animals, or things) to generate and process data, solve problems, or inspect phenomena. For these reasons, although the terms "human computation games" and "citizen science games" are more precise than "crowdsourcing," they may not be inclusive enough for the types of games that this book discusses.

We can broaden the term "citizen science games" by shortening it to "citizen games." A benefit of this term is that it suggests openness of participation—as everyone is a "citizen" of somewhere—while still capitalizing on the knowledge-focused, participatory "citizen science" history. However, "citizen" by itself suggests games with a civic, social studies, or political purpose—such as ones related directly to developing citizenship skills or being a good citizen—and also may imply that players need to be legal citizens of a particular region, rather than, more generally, active, globally engaged citizens of the world. Moreover, the term "citizen science" has not achieved as much permeation, popularly, as the term "crowdsourcing" or even "human computation."

Thus, naming this type of games is no easy feat. More general names seem too shallow, trendy names too inexact, and more specific ones too exclusive. We do not know how these games will continue to evolve in the future or how they will fit in societally, so it's hard to predict their common usage. We could call them problem-solving games, as that is what they ultimately aim to do, and indeed the term "applied problem-solving games" has been used. But *all* games involve some type of problem solving (even if it's just to solve the problem of completing the game), so this may be too broad. "Contribution games," "crowd games," "collective games," "participatory games," and "human games" are all too vague, as all games require some type of contribution or participation; not all of these games may invite human beings, or even collective or crowd action; and some may just invite experts, individuals, animals, or a select group of stakeholders. We could call them "change games," but that gets confused with Games for Change, a movement, organization, and category of games comprising games that express or encourage social change. We can also argue that all games have a purpose or desire to make change for someone, whether it is to increase profit, enhance enjoyment and entertainment, or solve humanity's mysteries.

Gordon, Walter, and Suarez use the term "engagement games," which they describe as a new type of serious game. They contend that engagement games enact real-world processes through the game itself and that they are "a new type of interface for real-world processes . . . from community planning and data collection, disaster preparedness, advocacy and fundraising, to skill and network building." The types of games described in this book fit the definition of engagement games because the ultimate goal of these games is to make real-world change. But the definition of engagement games seems to encompass many more forms of social action, not just knowledge-building and problem solving.[23] Also, all games involve engagement, whether engaging the player with the game or the game with the outside world, so this label may not work. Likewise, in her book *Reality is Broken*, Jane McGonigal uses the term "social participation" games in a similar way as Gordon, Walter, and Suarez apply "engagement games." Again, while McGonigal's definition suggests some type shared or collective activity, the definition may be too broad in that it describes games that have any type of real-world action or counterpart.[24] In addition, many games invite social interaction and participation, so applying this term does not necessarily distinguish these games from other types.

The term "innovation games" might work. Innovation suggests change, growth, and contribution to the body of knowledge of the world. Shaffer defines innovation as that which *"cannot be standardized,"* and he explains that it is a practice or social process that happens in tandem with people and communities (such as other's works, ideas, or input), almost as if it is a conversation with the world. In other words, during the process of innovation, we feel as though others are watching, judging, and interacting with us, either implicitly or explicitly, even if those evaluations are not directly expressed, in a similar way as when we play.[25] Innovation could involve additions to scientific or social scientific knowledge, alterations in how we perceive humanity, theoretical contributions, or physical or engineered changes to the world. It is broad enough to include many different types and functions of games and game players but also specific enough to relate to the creation and expansion of new knowledge without being didactic.

But is "innovation" the term we are searching for? There is a mythology around innovation. Applying the term "innovation games" may hint that games should continue to outpace themselves, particularly as we produce more advances and knowledge; more effective analytical or design methodologies; sharper techniques for storing, analyzing, and using data; and more engaged communities—and as we are able to better motivate and include more people in the process.

The term also suggests that games will always lead to progress for the better. We often believe that innovation is always good—that it leads to growth, better outcomes, living standard increases, and more societal benefits. However, it may not even be as useful to innovate, per se. Strumsky, Lobo, and Tainter argue that it may be more effective to copy or imitate than to innovate when it comes to producing the new knowledge that helps our society and uncovers solutions to both common and complex conundrums.[26]

Therefore, for the purposes of brevity, inclusiveness, and specificity, I will call the games discussed in this book "knowledge games" because they seek to produce knowledge; solve authentic, applicable problems; or generate new ideas and possibilities for real-world change. This name clarifies the general goal of these types of games—to create knowledge that we did not have before and solve problems to which we did not previously have solutions—but it does not dictate the process, topic, design, or people involved with the game. "Knowledge games" overlaps with other terms, such as educational and learning games, but it suggests that the outcome of the game is knowledge, rather than just learning and education. As of August 2015, no individual, team, or institution currently labels its games as "knowledge games." It is not a perfect name, but I believe that, for now, it best encapsulates the types of games described and addressed in this book.

When I use the term "knowledge games," I am referencing the set of practices, contexts, designs, and relationships that emerge from and around those games with a goal or sub-goal of generating new knowledge about humanity, society, the universe, and any previously unknown phenomena. I am analyzing, discussing, and dissecting those games that invite human beings or other entities to play and work collectively (either with others or individually, or in tandem with computers or other nonhuman entities) to contribute information or to perform tasks or activities that can help answer specific questions, create new knowledge, and contribute insight to our understanding of the world. Although knowledge game players are playing and problem solving within a game environment, the knowledge created has implications for real-world processes, policies, problems, and people.

But can any game—and not just knowledge games—give us insight into humanity and our world? Just like books, film, sculpture, and other forms of media or art, can games also inspire real-world change? For example, can we learn about ourselves through the way we design a game or the experience of playing it? Or, are knowledge games uniquely poised to spur problem solving and insight?

If all games do provide us a window into humanity, it is no surprise that many researchers are using games as sites in which to study people and phenomena, just as they would use any laboratory or naturalistic setting. For instance, Lofgren and Fefferman used *World of Warcraft* (*WoW*) to better model responses to real-world biological outbreaks. They looked at the accidental unleashing of an extremely strong digital monster in *WoW*, which swiftly wiped out many avatars with lower HPs (hit points), to understand the responses of the players to this virtual epidemic and apply this to real-world crises. The researchers discovered, among other things, that they needed to account for journalists (the ones who ran toward the crisis) in their epidemiological model.[27]

To clarify, while all games may be able to contribute knowledge of some type, knowledge games in particular are designed with the primary intention of knowledge production, though other goals, such as entertainment, may also be salient. Using *WoW* to study phenomena does not transform it into a knowledge game. On the other hand, creating a new game specifically to study that phenomena could make that game a knowledge game.

Finally, I am not arguing that all games should be knowledge games or that all research should be done through games. Knowledge games may be another way to learn about the world, but they do not replace or necessarily surpass other knowledge-seeking methodologies or experiences. We should still be learning about the world through many different lenses—whether using traditional research methodologies, games, or future ways that have yet to be designed or imagined.

Categorizing the Contributions

Despite a lack of clarity about which label to use when creating, considering, and using knowledge games, we should still be learning from crowdsourcing, citizen science, and other contribution-culling phenomena. First and foremost, what are the types and categories of activities and processes being contributed, and could we apply these appropriately to knowledge games?

Let us examine two typologies from crowdsourcing. The first is described in Jeff Howe's 2008 book, and his categories are (1) crowd wisdom, (2) crowd creation, (3) crowd voting, and (4) crowd funding. Brabham critiques this division as concentrating too much on the function and process of the crowdsourcing, rather than the goals of crowdsourcing. Brabham instead creates four approaches to crowdsourcing activity that focus on the types of problems crowdsourcing seeks to solve, from the perspective of the organization creating and inviting the crowdsourcing platform.[28]

1. *Knowledge discovery and management*: In this approach, there is an assumption that some type of knowledge already exists (e.g., a pothole on a village road), and it is up to the crowd to be able to bring it forward, so that everyone can benefit from it. There is a collective commons where this information is made public or accessible so that it can be acted on. Brabham uses SeeClickFix and Peer to Patent as examples of this approach.[29]

2. *Broadcast search*: An organization believes that someone or some group has the ability to solve an open problem or contribute a useful idea or innovation. Brabham uses InnoCentive as an example.

3. *Peer-vetted creative production*: The purpose here is to invite the public to create, design, or curate new ideas, objects, or other products, as in Threadless.

4. *Distributed human intelligence tasking*: This approach invites the crowd to do tasks, analyses, or data collection, particularly on assignments where human beings are better than computers. Brabham uses Amazon's Mechanical Turk as an example.[30]

A concern with Brabham's typology is that it sometimes conflates the goals or problems that are trying to be solved using crowdsourcing with the way that the organization or the public participates. For example, the first type, knowledge discovery and management, is a goal (and a vague one), but the second type, broadcast search, seems to be a strategy that an organization might use to bring out solutions or new knowledge. The goal in broadcast search is really for the public to help solve problems, which is true about all four categories. Similarly, the distributed-human-intelligence approach describes tasks that human beings are better at than computers—but this is true also about the examples included in other categories, such as creating new T-shirts with Threadless or reporting problems in one's community with SeeClickFix—so the typology becomes tangled.

We can also look at a third typology, which is based on citizen science projects and was developed by Wiggins and Crowston, two researchers at Syracuse University's iSchool. They criticize earlier citizen science typologies as focusing on how the public participates in aspects of scientific research, rather than the way that that participation is designed or the "sociotechnical and macrostructural factors influencing the design of the study or management of participation." For example, those earlier typologies might include common categories of tasks, such as data collection. Instead, Wiggins and Crowston categorize citizen science projects by goals (such as science, education, or conservation) and consider how these different goals interact to create clusters of projects with multiple goals. Their first cluster involves all of the goals fairly equally, whereas their second cluster rates science extremely high as a goal, and less with management, restoration, or action. The goals that emerged from their analysis are: science, management, action, education, conservation, monitoring, restoration, outreach, stewardship, and discovery.[31]

Toward a Typology of Knowledge Games

Likewise, there are many different ways we could categorize knowledge games. We could divide them by the type of knowledge created and use disciplinary boundaries, such as scientific or STEM, sociopolitical or social scientific, humanistic, and artistic or aesthetic knowledge. However, these disciplinary borders are artificial. Most mysteries and problems require the overlapping consideration of many disciplines and lines of inquiry—from design and technology to human and social factors to scientific understanding. For example, designing a new healthcare system could fit across all of the aforementioned categories. In fact, maintaining one disciplinary lens could be a reason why a problem is still unsolved.[32]

We could instead divide knowledge games by the type of organization trying to produce new knowledge through a game. This could include categories such as social, individual, scientific, governmental, healthcare, international/global, corporation, nongovernmental organization, or not-for-profit. We could categorize knowledge games by the scope of the problem or sphere of the knowledge being produced, such as national, local/community, global, or multinational; or by the specific research or project goals, such as conservation or discovery. We could classify by the main purpose of the organization, such as outreach, community-building, or education. Again, however, for all of these there are overlaps, and these divisions may not be useful. How do we even, for example, judge something as local when knowledge created for local purposes could have wider and broader international implications? Does the purpose of the organization matter or are the game's goals more salient when categorizing knowledge games?

Another option might be to organize knowledge games by the type of gaming platform being used (e.g., online/browser-based, social media / Facebook, or mobile), genre of game (e.g., puzzle, platformer, role-playing game [RPG]), or type of game mechanics used (e.g., finding, sorting, resource management, jumping, planning, or socializing). We could divide by the design principles that drive the game (e.g., using "jigsaw" arrangements to spur collaboration or motivating through the use of story) or issues that emerge from the use of the game (e.g., accuracy, privacy, bias, or recruitment).

Alternately, we could look at the knowledge games that have been created and see what types of design patterns or clusters emerge. The problem with this and all of these possible categorization schemes is that there are very few knowledge games that currently exist, so attempting to classify them is challenging.

For the purposes of starting the conversation, I created a tentative typology for knowledge games, with consideration to the previously described citizen science and crowdsourcing typologies and to existing knowledge games. This typology focuses on the primary goals and functions related to knowledge production that designers and investigators may seek to solve through knowledge games. This typology should be vetted empirically as more knowledge games emerge, but it should not drive their creation. Rather, this categorization scheme should remain flexible and open to novel ways of designing and using knowledge games. Just as new genres of games continue to emerge (e.g., MOBA, alternate histories, cinematic interactive fiction), new knowledge game formulations should as well.[33] In fact, many knowledge games may stretch across multiple categories, and each category is not necessarily meant to be mutually exclusive. Also, one category is not better than another, and which one works best for a particular knowledge game depends on the unique goals and needs of a project (see appendix A for a chart of the proposed typology).

1. *Cooperative contribution games*: These games invite players to contribute some type of activity or task, such as the processing of images or text, recording data from one's environment, brainstorming ideas, or categorizing and identifying objects or items provided through the game. Most of the time these games are cooperative in nature, in that people are working individually toward a common goal, though they could have more directly collaborative aspects to them. Typically, there are "right" or "more correct" answers that can be provided by the public, even if those answers are currently unknown. For example, when categorizing a moth shape or cell color, there are answers that are more accurate or would be more likely given by a majority of people. Or, for example, a game might use a needle in the haystack approach and search for that one accurate answer amid the cacophony, with the thought that the right one could be uncovered by using a large enough crowd. This is the category currently with the most games—and it is a category that may become more splintered in the future as more subtypes emerge. Current examples include *Happy Moths*, the *Citizen Sort* collection, and *Reverse the Odds*.[34]

2. *Analysis distribution games*: In these games, the player is not just providing, collecting, or processing data but also providing her own unique perspectives, strategies, interpretation, or perceptions on the collected data. How people interpret this information may help us understand more about the myriad ways people think, feel, live, and learn. There are not necessarily more right or wrong answers associated with the objects or information analyzed through these games, though accuracy or comprehension (at least in participants understanding what is asked of them) is still important. Examples include *VerbCorner, Who Is the Most Famous?, IgnoreThat!*, and *Apetopia*.[35]

3. *Algorithm construction games*: In these games, players engage in complex interactions, such as manipulating proteins, sharing their

dialects, or performing a schema in a virtual restaurant, so that a computer (and the investigators) can learn and shape an understanding of the phenomenon. Or people may be describing steps of a process, or testing others' designs, so that a computer could process this information. Often, these games help create, for example, an algorithm database to use in understanding behavior and/or predicting future behaviors. Examples include *The Restaurant Game, Foldit, EteRNA, The SUDAN Game,* and *Which English?*

4. *Adaptive-predictive games*: These games take the information, interpretations, and algorithms generated (as in the other categories); model this information and/or create new algorithms that can make predictions on anything from behaviors to attitudes to game play; and then reshape themselves and adapt to the new information. This type of game also may change in real time or for subsequent players and play-throughs in an effort to evoke real social, individual, or scientific changes. Straightforward adaptive games such as *FASTT Math*, as well as more complex ones like *Nevermind*, presage this, but these games are learning and adapting based on individual interactions with the game rather than making predictions based on models molded from numerous people.[36] Games such as *SchoolLife* may approach this label, but otherwise there are no known knowledge games like this at the time. I believe this is the next frontier of knowledge games.

In the next chapter, I provide a more comprehensive introduction to the "game" aspect of knowledge games. What are the general design principles, research findings, and frameworks that support the use of games for knowledge production and problem solving? How can design help ground us as we explore the key questions and concerns of knowledge games and meet their potential?

CHAPTER TWO

Design

Click. Swipe. Zoom.

I take out my iPhone and help care for the Odds, little gray creatures who need their world rebuilt and replenished. In *Reverse the Odds*, a mobile game, I solve mini-puzzles that help me "re-colorize" the dreary world of the Odds by earning upgrades to their *Q*bert*-like platforms. I use special potions to help me solve the puzzles, but these potions run out quickly—and you need to wait two long minutes to replenish one at a time.

For the next hour, I continue to solve puzzles to improve the land of the Odds—after all, there are 175 levels of gameplay to complete—and I want to succeed in the game and help the Odds. As in other social games, like *Cookie Jam, Hay Day,* or *Candy Crush Soda,* I want to keep solving puzzles, completing game tasks, and earning those potions.

It sounds like just another addictive free-to-play game, right? But it's okay. I'm spending my time playing a mobile puzzle game that helps Cancer Research UK discover more about lung, bladder, head, and neck cancer. *Reverse the Odds* was created by Channel 4 and Maverick Television in the UK and the Cancer Research UK organization, using Zooniverse, a platform for citizen-science-type projects managed by the Citizen Science Alliance.[1] In *Reverse the Odds,* you solve puzzles for the Odds but also look at samples of actual tumor tissue and make judgments about them.

To earn new potions in *Reverse the Odds* either you can wait or you can travel in-game to a virtual lab and use a cartoon microscope to inspect slides with cells on them. If you choose the microscope, what pops up next is an actual image from a database of tissue-sample images. For each image, you need to answer a series of questions about the types of cells present, such as color, brightness, and which types of other colors surround them. The game then compares your answers to other players' and gives feedback on when your answers are different from other game players'. So instead of using in-game tokens or coins (which can typically be purchased with real money), such as in other free-to-play social mobile games, you can spend time inspecting cell images to earn more potions. It's win-win, right? We get to play a fun puzzle game, while the massive backlog of Can-

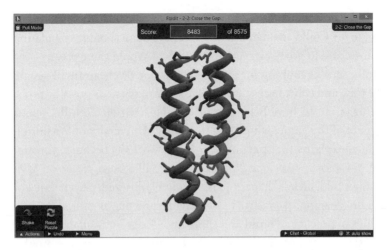

In *Foldit*, players fold 3-D representations of proteins to help us better understand possible protein structures. Source: Zoran Popović, University of Washington Center for Game Science

cer Research UK's images gets processed and our knowledge of cancer increases.

Likewise, in a now-classic knowledge game called *Foldit*, the puzzles *are* the knowledge making.[2] In this game, players work together and individually to bend and twist simulated 3-D models of proteins to try to guess their structure. *Foldit* was created in 2008 by a group of University of Washington scientists who initially wondered whether a game could enable participants to predict a protein's structure by folding and manipulating a 3-D simulation of it. They considered how computers are quite good at doing routine tasks, such as simple calculations, very quickly but are not so great at efficiently doing the types of tasks associated with protein folding. Human beings, by contrast, through intuition, spatial reasoning, and trial and error are actually better at the types of tasks associated with predicting protein folds. So the scientists conjectured that instead of having one group of lab-mates sit around and play with simulated 3-D models of proteins all day, why not have hundreds or even thousands of people try to predict protein structures and then codify their strategies via the Rosetta@home program so that computers could use it and learn from it. After all, there are over 100,000 proteins in the human body alone, and trillions of different possible orientations for protein structures to take.[3] One lab of even 20 people would not be able to put a dent in the problem, but over 400,000 active game players certainly could.

When *Foldit* first came out, it was not clear if it could really achieve what it set out to do. Could amateur scientists learn enough about manipulating proteins to be able to predict structures like a pro? Would game players—a group typically seen as looking for a frivolous escape or cheap thrill—really put in the time and effort necessary to solve some of science's pressing problems?

It turns out to have been a good gamble. Adrien Treuille, one of the designers of the game, who was a postdoc at University of Washington at the time, explains that part of why *Foldit* works is because, through the game, the proteins became "toy-like"—more like Lego pieces or *Minecraft* parts than cold, lifeless shapes.[4] In the *Foldit* playground, players can work together on meaningful constructivist activities and may organically form communities of practice to build and share knowledge and collaboratively solve authentic, situated problems using the real tools that a scientist might use.[5] Michael Nielsen, in *Reinventing Discovery*, explains, "Some of the controls built into the game are similar to the tools used by professional biochemists. The lower the energy of the shape the player comes up with, the higher their score, and so the highest scoring shapes are good candidates for the real shape of the protein."[6] This also means that from a design perspective, a high score or success in *Foldit* is intertwined with the success in the real-world problem solving of protein structures, potentially motivating success in both.

Thus, through *Foldit*, amateurs are able to make meaningful contributions to science by simulating the activities of actual scientists, using the realistic tools, problems, approaches, and epistemic frames that they would normally use—and through a game. After three years of successful test cases, in which players accurately solved protein folds that were already known, the *Foldit* players made a novel discovery. In 2011 the *Foldit* principal investigator, Zoran Popović, his team at University of Washington, and a team of *Foldit* players, published in *Nature* that they had figured out the crystal structure of a retroviral protease, a type of enzyme, called Mason-Pfizer monkey virus, which is implicated in HIV and AIDS. Understanding its structure helps in the creation of drugs, which can potentially help manage or control HIV infection.[7]

Toward Knowledge of Knowledge Games

So are we really helping to cure cancer and AIDS through games? Who would have thought, even just a decade ago, that games could contribute to our knowledge of two of the most deadly, devastating illnesses of our time.

In this chapter, I introduce the peculiar possibility that games and gaming experiences may be directly designed to help solve real-world problems and contribute knowledge—not only scientific and health-related knowledge but also knowledge more related to humanistic, cultural, social, and psychological questions, as well as to questions that span all perspectives.

Games, regardless of whether we conceive them as a set of practices, designed worlds, or type of participation, culture, medium, communication platform, or interactive activity, may be particularly adept, appropriate, or amenable to supporting problem solving and contribution to knowledge— and they also may be quite limited in reach. What types of research, frameworks, and theories support (and do not support) the place of games in knowledge production? What information do we still need to gather to better design and use knowledge games—what are the many gaps? What recommendations emerge from what we know so far about games, knowledge, design, human beings, computers, culture, and current research?

In this chapter, I map out and summarize the relevant topics and research related to games and game design, which may help us better understand the potential and limits of games as effective knowledge-producing experiences. Each of these topics is examined in much more depth in subsequent chapters, and I continue to dissect and expand on these possible benefits and pitfalls throughout the book.

Problem Solving

Games at their core involve problems or goals (e.g., find a way to the flag in *Super Mario Brothers*), which players need to solve or complete. There are rules and constraints in a game's world and a realm of lifts and levers that can be pushed and pulled so that the player can tumble toward each problem's solution. The game possibly even serves as a physical or virtual manifestation of a problem's "problem space."[8] Games may be particularly effective for practicing problem solving because they can enable players to feel and observe cause and effect and choices and consequences just by experiencing what is and is not allowed, or what is rewarded or punished, in the game's world. Players can test out hypotheses about how to solve a game's problems (e.g., how to overcome an obstacle), such as through trial and error, educated strategies, or modeled behavior from other players. Moreover, games have been credited in honing players' problem-solving skills, such as through encouraging alternate paths, providing stories or analogies, or encouraging the sharing and deliberation of different perspectives among players.

Can games help people not only to become better problem solvers but to actually solve real-world problems? Games are not necessarily appropriately designed for all types of problem solving. In chapter 3, I discuss problem solving, which types of real-world problems might be better solved through games (as opposed to other methods), and how games might (or might not) enhance the problem-solving skills of their players and help solve problems in particularly effective ways.

Motivation

Scientists, social scientists, and humanists who want to use games to better understand how people think or feel, or who want to enable people to contribute to a data-collection or analysis initiative, consequently need people to play those games. Analogously, just as any researcher leading a research study or a data-collection project needs to recruit participants, those creating games for these purposes also need to recruit and motivate participants. Games, if well designed, can be particularly motivating for an appropriate audience, which make them potentially useful for activating a large number of people. However, potential players are being pulled in many different directions—by work, family, and other pursuits—while being tempted by more mainstream games. How do knowledge game creators entice someone to play their games rather than the latest *FIFA*, *Grand Theft Auto*, or *Call of Duty*, or even *Gone Home*, particularly when they may not have the budget, resources, time, or distribution abilities to do so? Should they even try to?

Moreover, knowledge games may not be as visually appealing, narratively engaging, or dynamically designed as their mainstream counterparts. Yet games that are for knowledge production may have other factors beyond just the game itself that can help motivate players, and understanding this full picture of motivation is important when creating and using them. The ways in which an audience is motivated to participate in games, citizen science, and crowdsourcing activities; volunteer work and leisure pursuits; and knowledge contribution and civic engagement in general, are explored further in chapter 4.

Collaboration and Cooperation

Not all problems or tasks are better done with other people (or even other devices, such as computers, or nonhumans), but when properly designed, games can be particularly adept at enabling social interaction and collaboration, as well as the individual actions that can support more collective goals. Many large-scale problems, for example, need people with different experiences and expertise working together collaboratively, or they may need people working individually so that they can cooperatively contribute to a shared repository or data set. The social interactions that occur within a game, or even around a game (such as in the case of players actively sharing the game with others through word of mouth or social media or by communicating hints, tips, social mores, and words of encouragement in forums) can also possibly motivate users. Problem solving, motivation, and social interaction can also work together in complex ways to make each component more effective, and greater than the sum of its parts. They can also work together in deleterious ways to discourage or demotivate. Likewise, not all games are adept at supporting social interaction, whether informal sharing outside of the game, or intense collaboration or competition within the game. In chapter 5 I discuss the intricacies of collaboration, cooperation, and social interaction; possible interconnections with problem solving and motivation; and when interpersonal interactions are more or less effective in helping to contribute knowledge through games. To reiterate, however, not all games for knowledge production need to encourage or rely on collaboration or other types of social interaction.

Participation

To some degree, games can be scalable and accessible, but they do not necessarily enable participation universally and equitably. Only a portion of the public can both access and participate in games, depending on the platform and technologies required. However, digital games do have the potential for supporting and including a broader range of people, for example, than a typical local research study (though this does not necessarily make games better avenues for knowledge production than traditional research methodologies). Games, which can be played almost anywhere, can enable more participants to work on a project than a research team could ever undertake by itself. However, these potentials are highly contingent on the types of questions and problems explored in such a game, and the types of distribution platforms or designs used, as not all games will be able to motivate and engage an appropriate or representative population. Games can also be intimidating or inaccessible for a variety of reasons, including technological, logistical, social, or psychological causes. Certain age, gender, ethnicity, national, and economic groups may be underrepresented in knowledge games. For example, what happens if a knowledge game can only appeal to traditional game players, rather than the general population, or is only able to be played by people with access to particular mobile devices or Internet-enabled computers? What happens if only those with enough leisure time are able to provide their play contribution? What are the implications of games that use "amateur" participants, and how does this affect the relationship between knowledge professionals and amateurs? Games do not necessarily invite all types of players, and game culture does not necessarily help all people feel like they are potential participants. Chapters 6 and 7 explore issues of participation, including amateur participation, the relationship between organizers and participants, the accessibility of games and game culture, and how this relates to the design of knowledge games.

Frivolous, Fun, and Engaging

By definition, games, if well designed, should be fun first and foremost—right? Yet some games may not even feel fun at all but instead seem tedious or time-consuming—not all games that purport to contribute knowledge are fun for everyone (or even anyone). Could the "fun" of games become downplayed in lieu of other purposes and goals? Moreover, the "fun" aspect of games, and the equation of gameplay time with leisure time, may problematize their seriousness and ability to contribute knowledge. If the focus in games is on fun and not work, can we take their outcomes seriously, and can we consider them rigorous enough? Does the fun supersede or exploit the oftentimes serious topics of these types of games, such as cancer, AIDS, or poverty?

On the other hand, we can argue that games are not just simply fun and frivolous pursuits but that they also involve and evoke work and labor. Players are not simply playing, but they may be play laboring. The boundaries between play and work in knowledge games may be blurred even more because the purpose of these games is to contribute knowledge. They may involve the tasks and activities that are normally conceived of as work, and not play, such as identifying and categorizing images or collecting specimens, further conflating work and play.

Knowledge is not devoid of social context. Games are often framed and discussed as being fun and frivolous, and this perception matters and may affect how the public accepts knowledge from them. Likewise, perceptions of social science and humanities research affect the public acceptance of their methodologies in terms of their validity and rigor—and further complicate the use of knowledge games for not directly scientific knowledge production. Managing misperceptions about games and social science and humanities research, so that the public wants to participate in, contribute to, and trust knowledge gained from games, is essential to the future of knowledge games. The tension between (and possible overlap of) fun and labor in knowledge games is discussed further in chapters 7 and 9, as well as the issue of knowledge production and acceptance.

Complex Systems and Interactions

Real-world problems and conundrums, whether solving the healthcare crisis, understanding depression, or enabling greater access to education, are highly complex. Healthcare is not just a simple concept we can figure out on paper but a living, breathing social, cultural, economic, political, and scientific system. Likewise, games are dynamic systems. Players enter and explore within a game world, which is a deliberately designed system based on a set of rules, assumptions, and values and is further problematized by the addition of other players and their own unique activities. While games cannot possibly simulate or model every corner of a dynamic system, they can realistically reimagine aspects of complex processes, skills, actions, and information within an authentic context, bringing people together over time and across formidable distances. Shaffer points out that games can potentially simulate epistemic knowledge, or knowledge that is based on realistic models or tools.[9] For example, participants in *Foldit* use authentic 3-D models of proteins, just as scientists would, and players in *Reverse the Odds* use real images of cancerous tumor cells to make judgments and classifications. Using authentic materials in games can enable more relevant contributions or analyses of data, whether people are playing with proteins, cells, moths, or galaxies.

Simultaneously, we are becoming more and more able to process, analyze, and generate powerful models based on large-scale data sets, which could inform in-game systems and make them more accurate and relevant, even potentially predictive. In chapter 8, I consider issues of designing game systems, particularly in regard to the accuracy and validity of data analysis methodologies and conceptions of data, and the concomitant issues of privacy and ethics. For instance, public interaction with connected platforms has always brought up questions of privacy, but working through the technical, logistical, theoretical, legal, and psychological aspects may also be particularly pressing with publicly accessible knowledge games.

Human and Humanity

Play and games are part of what it means, and has always meant, to be human, and even nonhuman entities play. Therefore, it may not be such a stretch to use games to learn more about humankind, to contribute to human knowledge about the world, and to better understand ourselves. After all, we could make the claim that all games, no matter what their purpose, through their design, play, and emergent culture have always taught us something new about ourselves and our world.

What are the implications for playing knowledge games for society, for humanity, and for knowledge creation itself? What are the epistemological changes necessary for knowledge production to come from, of all things, games? How does the existence of knowledge games shape the institution of (and institutions involved with) knowledge, and how are these games themselves shaped by the systems of knowledge creation? Does this shift what it means to be engaged citizens and global residents? Do games enable a more democratic approach to knowledge creation and sharing, or do they replicate the hegemonic forces that already exist? Knowledge production and its connection to humanity, what it means to be human, and what it means for games are discussed in chapter 9.

A Design Approach

For a week in late February 2015, an image of a dress went viral. The question was simple: is the dress white and gold, or is it blue and black? It seems like there should be a clear-cut answer, but the truth has been elusive. Everyday citizens, politicians, and even celebrities (such as Taylor Swift and Mindy Kaling) were weighing in through tweets and Facebook posts, or through conversations with their friends—and both sides were vociferous in their opinion. Even professors were analyzing why some people saw one set of colors versus another, explaining that variances in the number of blue-sensitive cones in one's retina, the modes of light to which people were typically exposed, or even differences in color perception (people perceiving the same color differently) may be partially responsible.[10] I, for one, saw the dress differently when I was and when I was not pregnant.

The premise behind the dress image—our differing perceptions of color—is also what seemed to inspire two games, *Specimen*, a mobile game, and *Apetopia*, a free, online, arcade-style endless runner game out of Germany. In both games, the player performs tasks related to color perception, which ostensibly reveals individual perceptual differences.[11] *Specimen* is not designed as a knowledge game, as the primary goal is entertainment according to the designers, but it certainly could be adapted to be one. *Apetopia*, however, is more readily labeled as a knowledge game, as its designers state that a primary goal is understanding color perception and individual differences. In *Specimen*, a player needs to tap the kidney-shaped specimens, shown in a cartoon petri dish, which best match the game's background color. Likewise, in *Apetopia*, a player gets a target background color and then needs to choose which color gate to enter based on which gate's color he thinks is most similar to the target color. For example, the target color might be a baby blue, and the gates might be turquoise and periwinkle. Both games move quickly, so the player needs to make snap decisions about which specimen to tap or which gate to enter.

Although both games have the same premise, their game designs differ. For instance, in *Apetopia*, the game takes place in a dark but beautiful post-apocalyptic, graveyard-like world, where the only colors that stand out amid the gray background are the target and gate colors. In between choosing the colored gates, the player must also avoid hitting bombs or trash and instead collect coins. Bombs and trash lower your resistance while coins bring your resistance up, so that you can spend more time running through the repetitive world. The farther you run, the more distance you cover, and the higher you end up on the game's leaderboard. Sometimes you get speed boosts, which help you cover more distance more quickly, but it was never clear to me why I received a boost. In one play-through of *Apetopia*, after a few minutes of choosing between colored gates, no more gates were presented to me, and I just ran through the repetitive world until I surrendered. *Specimen* uses minimalistic art. The game's interactions take place in a cartoon petri dish, from which you can view a changing variety of colored "specimens." While choosing the specimens that best match the background of the game, you can get extra time boosts by tapping on patterned specimens. Whereas in *Apetopia* your goal is to gain the most distance, in *Specimen* you are racing the clock and looking for ways to extend your time.

Let's dive into *Apetopia* further. Unlike some knowledge games, there is no tutorial level or embedded tutorial in *Apetopia*. Instead, you are thrown into the full game immediately, with no explicit explanation of the rules except for the game's brief introductory screen. Normally this is sufficient because a good game should provide you with immediate and appropriate feedback on how you are interacting within the game's constraints so you

In *Apetopia*, an endless runner game, players choose from among two colored doors to provide insight into color perception. Source: Dr. Kai Uwe Barthel

can understand its system. In other words, you learn what is appropriate or inappropriate, rewarded or penalized in that system by playing in and experiencing that system. In *Apetopia*, however, while you do get some negative feedback from hitting obstacles and positive feedback from hitting coins and landing on the leaderboard, the game often moves too quickly for players to understand the rationale behind their successes or failures.

Apetopia has no reward for choosing one colored gate versus another—which is appropriate as this would bias the goal of the game (which is to understand but not reward particular color perceptions). However, it also means that the game ends up being more about avoiding obstacles, gaining distance, and wondering whether different gates help you differently, rather than about judging colors. Moreover, the game is repetitive. After a while, it feels like any gate decision is arbitrary, which means there is less motivation to choose either gate, and, I argue, it makes your choices in the game feel less meaningful.

Apetopia and *Specimen* illustrate why design matters so much. First, although their central theme (color perception) is quite similar, the execution of the two games is quite different. Second, *Apetopia* is an example for why the integration of game mechanics and project goals are important, yet so challenging. The process of discerning colors in *Apetopia* could be integrated into the gameplay rather than tacked onto the "real" gameplay (e.g., avoiding obstacles and gaining distance). At the same time, I understand why *Apetopia* does not reward players for choosing the "right" color door, as there are no right answers. Some people see a black and blue dress and some a white and gold dress—and both perceptions are essentially "right." This is a key design challenge for knowledge games—how do we reward correct or appropriate solutions when we do not yet have the answers or none exist?

Nevertheless, *Specimen* and *Apetopia* may enable us to better understand human color perception.[12] But despite the usefulness of any knowledge we gain from a game, that game is not necessarily well designed, fun, or even fully effective. In the previous section I listed seven possible issues related to games and knowledge production. In this section, I further consider the process of design. How do we translate the needs and goals of a project into a real knowledge game, while also considering those seven issues? How do we ensure that the would-be knowledge-making possibilities of games are realized? This is where design comes in.

First off, what do I mean by design? Design often gets conflated with other terms. For example, people may focus on design as art creation or as the artistic or aesthetic aspects of a particular artifact, such as a website's colors or images, or a game's art style. Or people may talk about specific types of design, such as product, fashion, or software design. In this book, I use "design" more broadly, conceiving it as an active process of finding the best solution to a given problem based on the available resources and constraints, designers' previous knowledge and experiences, and the desired outcomes or goals. Thus, design could be seen more as problem solving, where a problem is stated (e.g., as in the case of the game *Apetopia*, a lack of knowledge about human color perception) and a particular solution is fashioned (a game is designed that invites players to choose between two gates, avoid obstacles, and find coins).

This vision of design suggests that it is a set of practices or lenses that can be applied to almost anything—an experience, a process, or an object.[13] For example, we can design a research protocol based on a stated research question and select from different methodologies to approach the question. Or we can design a pen or a chair from a set of specifications, audience needs, budgetary constraints, and manufacturing abilities.

The solution to the stated problem (e.g., the mysteries of color perception) could have been conceived an infinite number of ways. Instead of ending up with the game *Apetopia*, we could have devised *Specimen*. Or the solution could have been a paper task, where a researcher shows printed color samples to participants and asks which color is most similar to a target one. Instead of an online, endless-runner-type game, the same color-perception task could have been integrated into a puzzle game or role-playing game. Success in an endless runner game could have been based just on number of gates chosen rather than on obstacles avoided or coins gained. Thus, the design grows out of answers to the many questions about how to make a game.

Just as with designing a research study, how you design a game to solve a particular problem will affect the results. If you opt to use a choice task rather than a focus group, a sample of color-blind participants to query versus a large group of college students with normal vision, it will also affect the type of color perception knowledge you contribute. Using a game rather than a paper survey or an in-person interview rather than a web-based interaction will alter the knowledge that is produced.

If a game is chosen as the best solution for a given research question, its design needs to be carefully considered and, likely, iterated with actual game players in various contexts to ensure it meets the goals, needs, and expectations of a project.[14] A poorly designed game may provide faulty or heavily biased results or may simply not motivate the appropriate sample of players you hope to reach.

Designing a knowledge game may even be more challenging than developing a typical game. Designers of these games need to weigh several concerns, such as accurately and appropriately representing an open problem, making a game engaging for the right players, and dealing with the requirements of scientist or researcher stakeholders as well as the desires and interests of potential players. Designers of knowledge games need to incorporate and weigh epistemological and theoretical concerns (such as the seven discussed previously in this chapter), as well as the pragmatic, logistical, artistic, experiential, and computational constraints of creating a real-world experience, to come up with their best guess at the right game. For example, it may be challenging to devise the actions and mechanics in knowledge games to match or complement the real-world problem-solving activity or the knowledge needs of a game. Designers may approach the same concerns very differently. In *Foldit*, game players actually fold simulated proteins to figure out their structure, which mimics its real-world counterparts. In *Reverse the Odds*, by contrast, players solve puzzles that are not related to cancer analysis but then look at cancer cells to do the analysis, earning objects that can be used in the core gameplay. Neither of these approaches is necessarily better or more effective—what works best for a knowledge game depends on the mix of target audience, research questions, and project needs.

We can imagine and theorize about why a game works and why a game fits our interests or connects with us, but, ultimately, we like what we like, and the only way to truly know if a game works is by testing it out, or playtesting it, with others.[15] How do we playtest a knowledge game? There are obviously some inherent challenges in playtesting these games. First, it is likely that if you are designing a knowledge game, you will go through many, many rounds of testing. During testing, you may not always want the extra burden of checking and double-checking the accuracy of any contributions. Instead, you may want to focus just on the game-play aspects, such as how people are interacting with each other; their engagement with the game, comprehension, and facility with the main actions or mechanics in the game; or their desire to continue playing the game in the future. Then, or separately, you can test the accuracy of contributions (particularly if you are creating a cooperative contribution type of knowledge game; see chapter 1), or you can investigate how people are engaging with the game's contributory aspects and how this may or may not affect the accuracy or quality of any contributions. Finally, you will also need to playtest all of the components together, making sure they are all in sync and ensuring that project goals match game mechanics, that the act of contributing makes sense within the context of the game, and that there is fluidity among all aspects of gameplay and knowledge production. For instance, the contributory aspects of providing color perceptions should not feel external to the game world or gameplay; rather, they should be relevant and meaningful within the context of a game.

Just having game mechanics and project goals match up may not be enough, however. You may want to ensure the authenticity and relevancy of those mechanics or actions in the game to the types of activities or tasks you want people to contribute. For example, *Play to Cure: Genes in Space*, a spaceship-traveling and resource-management game, has nothing superficially to do with cancer data analysis, even if the action of strategizing the route of the spaceship helps reveal cancer data discrepancies (see chapter 8). This is not necessarily a poor design, but it may affect the data you collect or whether the participants are motivated to play it, as well as how the game is received.

Thus, this book takes a design approach, and it considers not just the relevant research and theoretical frameworks but also how to translate this into actionable design recommendations. In service of this, I propose guiding questions and design principles for the interdisciplinary creation and use of knowledge games, which are culled from research and frameworks, and are explicated in the appendix.[16]

Finally, I acknowledge that the design and use of knowledge games are in their nascent stages. Although I am critical of the games and other media I discuss, I want to reiterate the challenges of creating these new types of experiences. Those who are designing and using knowledge games are pioneers, navigating a brand new field without clear maps and with limited resources. I respect how they balance an enormous number of concerns and constituents to create games that help us learn more about ourselves, and how each new knowledge game gives us insight into the possibilities of designing its future.

Defining Games

You may have noticed that I have yet to define games. The omission has been intentional, as knowledge games may help us to reimagine the role, purpose, and contour of games.

Jesper Juul, in *Half-Real: Video Games between Real Rules and Fictional Worlds*, defines games as having a "rule-based formal system; with variable and quantifiable outcomes; where different outcomes are assigned different values; where the player exerts effort in order to influence the outcome; the player feels emotionally attached to the outcome; and the consequences of the activity are optional and negotiable." Juul considers several other definitions, such as one from Johan Huizinga, a play theorist from the 1950s, who explained that games happen outside of "'ordinary' life . . . an activity connected with no material interest, and no profit can be gained by it." Likewise, Roger Caillois, in 1961, explained that play was "free (voluntary), separate [in time and space], uncertain, unproductive, governed by rules, make-believe." Notably, many of these older definitions of games refer to games as being separate—from everyday life, from work, from serious action, and from the productive sphere.[17]

But clearly *Foldit*, *Apetopia*, *Reverse the Odds*, and any other game that supports knowledge creation could be considered productive, serious, useful to the real world, and perhaps even work. Juul explains that games are "half-real," in that they cross real and fictional worlds, and that "to play a video game is therefore to interact with real rules while imagining a fictional world." Games have "negotiable consequences"—in other words, a given game "can be played with or without real-life consequences." He uses as an example a game like poker, which has betting; sometimes it can be played "for fun" and other times it can be played with money or professionally without diluting its meaning as a game.[18]

But what if the whole purpose of a game is to alter our understanding and change the world at large? If we are seeking to solve serious issues through games, and the very playing of a game is thus seriously productive, is it still a game? And if we, the players, create knowledge outside of a game, does that lessen its gameness? If games are knowledge makers, do they become redefined?

As we delve deeper, we realize that these questions force us not just to redefine games, their design, and their function in our society but to reevaluate knowledge creation, the practice of research, and how we know what we know—to rethink our entire epistemological frameworks. Could knowledge games help us know our world but also restructure, and even redesign, *how* we know it? These questions are explored in the rest of the book.

In the next chapter, I introduce types of problems, and investigate the potentials and limits of games in supporting problem-solving processes.

PART TWO

WHY KNOWLEDGE GAMES?

CHAPTER THREE

Problem Solving

Have you ever wanted to design brand new RNA (ribonucleic acid) molecules? In *EteRNA*, you can. In this knowledge game created by some of the original *Foldit* creators, you can solve open RNA-design problems disguised as missions and submit possible solutions to create a "large-scale library of synthetic RNA designs." When you begin the game, you are given a series of nucleotides, and you need to figure out where and how to connect them to make the best, most stable, RNA molecule.[1] You can manipulate the molecule, sometimes changing out nucleotides, and make hypotheses about which pairs should connect and which nucleotides should get replaced. Through trial and error, or educated guesses, you can investigate the consequences of your hypotheses and then decide on the next steps. Thus, at its core, *EteRNA* invites people to solve problems.

Once you complete the tutorial missions, *EteRNA* provides you with real RNA-synthesis missions, organized by difficulty level.[2] You can earn different amounts of "money" (in-game currency) and see how many people have previously solved the same problem, which is often in the tens of thousands. Each mission provides you with a series of specifications, or constraints, to which your RNA design must adhere. In other words, using these missions, the game crowdsources designs for RNA molecules and adds them to a database of RNA blueprints.

Once outside of the tutorial phase, the RNA puzzles vary widely. You can enter into the various mission "rooms," fashion a design for a new RNA molecule to meet given constraints, and use a chat function to communicate with other players. You also gain points for any solid solutions you pose, earn badges, and can land on the leaderboard. Thus far, over 37,000 players of *EteRNA* have synthesized hundreds of new RNA designs, creating a database of "nearly 100,000 data points," which has resulted in the creation of an automated algorithm, EteRNABot. EteRNABot, together with the human community, is better able to devise new RNA molecules than any other algorithms that currently exist.[3]

Complex algorithms, or steps to a solution, can also be applied to other types of problems. A group of researchers from the University of Southern

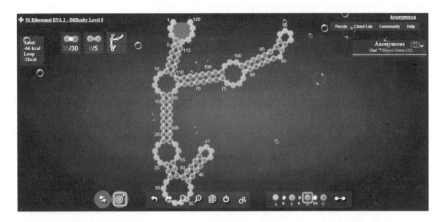

EteRNA is a game where players can design new RNA molecules based on given constraints. Source: Computer Science Department, Carnegie Mellon University

California's GamePipe Lab and Carnegie Mellon are creating a massively multiplayer online game (MMOG) that invites participants to simulate the steps to peace in the Sudan. In *The SUDAN Game*, participants test out various sequence of events that could lead to peace among the tribes. Based on the enormous number of tribal perspectives and scenarios, there are 185,760 possible interventions that need to be tested to evaluate if they lead to a peaceful outcome.[4] Having multiple people play through these sequences may help uncover the best possible algorithm for peace in the Sudan (at least based on the game's variables and constraints).

The intersection of games and problem solving suggests many questions. Can problems be more easily solved by everyday game players? Are games particularly adept at presenting problems and enabling steps toward their solution? Are there specific ways of designing and using games that better support problem solving? What types of problems are best served by games? This chapter explores these questions.

Human Problem Solvers

Problem solving is essential to what it means to be human. We are always solving problems, whether it is figuring out where our toddler hid her favorite stuffed animal, which route to take to avoid traffic, how to respond to a bully, or how to stretch our salary to buy enough food for the week. There are many different types of problems—personal, global, complex, and simple. Based on the need for problem solving in our lives and for our society, it is not surprising that finding ways to support problem solving is imperative.[5] The question is how? Could games help us solve problems? Could large-scale complex problems that are unsolved and ever evolving, such as global warming, curing diseases, reducing suicide and depression, or providing clean water, food, and medicine to all the people in the world, be better (or at least differently) approached through games? Could even smaller-scale, more personal, and individual problems be solved through games?

But first, what is a problem? A problem has two main, defining characteristics. The first characteristic of a problem, as David Jonassen explains, is that it "is an unknown entity in some situation (the difference between a goal state and a current state)."[6] So a problem is something as yet unknown, or for which there is no knowledge yet. For example, this could be a simple arithmetic problem or a complex social problem such as how to ask your boss for a raise. Or it could be an open question such as how to design an RNA molecule.

The second characteristic of a problem, explains Jonassen, is that the problem's solution must have "some social, cultural, or intellectual value" such that understanding or creating this new knowledge would matter. In other words, a problem is only a problem if someone cares about it.[7] Problems matter because we say they matter and agree that they matter; the presence of problems is socially constructed. There may be infinite would-be problems that exist but only when someone identifies, acknowledges, and defines it as such and wants it to be changed, does something become established as a problem.

Problem solving, thus, is the active process of finding a solution to a problem that has been deemed a problem. Researchers use the term "problem space" to explain the "mental representation of the situation in the world" or the mental play area of the problem. This problem space could be externalized (in whole or in part) to live within a game space or virtual game world; it could encompass both the real world and game worlds. While the problem space can be realized, in part, within a game, Jonassen explains that the "mental construction of the problem space," whether within the mind of one individual game player or shared across many game players, is the most critical aspect of the problem-solving process. To solve a problem, we must be able to manipulate this problem space, whether it is just a mental representation or additionally digitally, virtually, or physically manifested, or any combination thereof.[8]

We can also conceive of the realm of the game (which could include not only the actual game, but also communication forums around it, and even the players' minds) as a problem space in which to address many problems, large or small. These problems need to get solved to reach the goal or goals of the game and successfully complete it. Doug Church argues that game worlds actually *rely* on problems, such as the challenges and obstacles that a player must overcome to reach a goal.

Although the ways to solving a game's problems, or its potential paths from the current state (the starting position of a game) to the goal state (the successful completion of a game), are typically known to the designer, they are not always initially apparent to the player. For example, finding a way from one room in *Portal* to another area is a problem that needs to be solved, and it could involve many mini-problems and tasks to reach the goal, which the player needs to discover through playing the game. In *Gone Home*, there might be multiple ways to explore the game's home, but each player needs to find his own solution and his own way toward the keys necessary to unlock parts of the home to progress the story. In this sense, the problem itself becomes an impetus for play; the problem gets solved to unveil the game, but it is also what drives the game, in part.

The designer knows, dictates, or at least plans for what the possible problem solutions will be. This is executed through the game's design itself—and not necessarily by some didactic exposition of the game's problems and solutions (though a design document may explicate this as well). Through the game's design, the creator poses the problem and structures the possible solutions, which the game player can then inhabit, practice, and perhaps, depending on the game, co-create, customize, or reconceptualize. New solutions—even ones the designer had not predicted—can continue to arise from emergent play. Moreover, in-game problems can sometimes be solved in different ways.[9] For example, in the *Mass Effect* series, a player's choices may direct her more toward a Paragon or Renegade path, but either path may lead to successfully completing the game. Likewise, part of the fun of *EteRNA* is trying to figure out different or the most effective ways to solve the problem of designing an optimal RNA molecule, which may be as yet unknown to the designer.

Raph Koster explains that one of the reasons games are so effective, engaging, and fun is *because* they pose problems for the player to solve. In some ways, games themselves are like problems waiting to be solved.[10] By playing the game and seeking progression toward its goal, the player shows that she finds the game and its problems meaningful. Conversely, problems also become meaningful through the act of playing the game. However, once there is nothing left to solve in a game, we may become bored and disinterested as the challenge of the game is no longer there.

Furthermore, the game itself is a solution (or one of many infinite potential solutions) to a stated problem. It is someone's or some team's proposed solution to the best game possible within given constraints (similar to an *EteRNA* RNA design). For instance, a game could be a solution to the stated desire for a particular audience to experience the aftermath of a nuclear apocalypse, such as in the *Fallout* series, or for elementary-school-aged kids to be able to learn and practice math facts (such as arithmetic and multiplication) in the case of *FASTT Math*. Any game that is created is one of many infinite solutions to a given problem, and its specific and unique design depends in part on the sociocultural context and perspectives of the team, time period, current gaming trends, prior expertise levels and talents of the creators, available budget and resources, other technological constraints, and so on.

Kurt Squire argues that a game, if well designed, should effectively provide the necessary information, context, factors, abilities, and properties so that the player can figure out what the solutions are to any given problem in the game.[11] Thus, games should present problems and obstacles that are ostensibly surmountable. What happens, though, with knowledge games? After all, knowledge games may pose problems that are as yet unsolvable. Does a problem need to ever be solved, or even be possibly solvable, to be appropriate for a game? Does a game need to be winnable or solvable to be a good, well-designed game?

These questions are relevant because knowledge games upend them. Games *can* pose problems with solutions that may be as yet unknown to the player, the designer, and anyone else. These problems may not just be how to complete the game successfully but problems with real-world counterparts, such as the ability to better understand galaxies, moths, or social behavior. Even with the resources, data, variables, and materials provided in the game, the problems they pose may remain unsolved, as not all knowledge game players will successfully reach a game's goals or help achieve the real-world goals. The criteria for a well-designed knowledge game might include providing effective pathways toward solutions, even if those solutions are never reached. In fact, solving a problem may not even be the ultimate goal; rather, the goal might be to stumble toward the solution and open up new problems. Ian Bogost argues that games are not necessarily about engaging or solving problems but "troubling the idea of solutions rather than leading us toward them." Are real-world problems ever fully solved, anyway? Bogost describes "the awesome hugeness of the world and its problems, as well as their solutions, always partial, always tentative, like a giant mountain peering through the fog, impossible."[12]

How is the design or experience of the game affected if most or all of its players cannot solve its puzzles or problems? Do knowledge games reinterpret the relationships between problems and games? Knowledge games may be a new way of conceiving what a "problem" is, in that the game becomes a shared problem space in which to solve a real-world conundrum, but they also represent a place of problem exploration, where we can ramble toward half-real resolutions and co-create new paths toward solution possibilities.[13]

Types of Problems

This brings us to the next question: what types of problems are potentially best solved through games?

As alluded to earlier, problems come in all shapes and sizes. Some are personal and well-contained, such as needing to find a ride to the airport. Some are highly complex and require people with specialist backgrounds to be able to solve them, such as developing a new after-school center to support literacy for kids and adults.

Many different types of problems exist, so how do we categorize them as a way of understanding which types best fit with knowledge games? One way is to do what Jonassen did and write down hundreds of problems to create a comprehensive taxonomy. He finds that, in general, problems vary on characteristics such as how structured they are, their complexity, how they are represented, and how abstract they are.[14] His eleven classifications are as follows. With each category, I have also provided examples of current knowledge games and citizen science projects to illustrate the range of approaches that games could take.

1. *Logical*: Logical problems are typically "abstract tests of reasoning that puzzle the learner."[15] Most knowledge games are not simple logic puzzles (such as arithmetic questions) since these answers are already known. However, logic puzzles may be one step in a larger problem that knowledge game players need to solve. For example, in *Phylo*, the player must make best guesses for how to match up or align a sequence

of RNA, which can help trace genetic diseases.[16] To do this, the first step is to know which nucleotides can align with other nucleotides, and then to use this to make a logical pairing, given a sequence of base pairs.

2. *Algorithmic*: Algorithmic puzzles have a series of steps that can be reproduced and followed to come up with a precise, finite, and predictive solution, such as an addition problem or calculation of speed.[17] Because the solution is already known, these types of problems may not have relevance to knowledge games. However, some knowledge games are used to help to create algorithms that a computer could process and follow to generate new knowledge more quickly. This is seen in both *Foldit* and *EteRNA*, in which players figure out optimal solutions to an open puzzle and then submit their solution of algorithmic steps to a database.

3. *Story*: Story problems use context and semantic understanding to help generate solutions. In the way that Jonassen uses the term, it typically applies to cases where the answer is already known—such as in a calculation problem (e.g., a train is going 60 mph and leaves the station at 2:00 p.m.; at what time will it arrive at the next station, 120 miles away?). Applying stories to problem solving in knowledge games is compelling for cases even where the solution is unknown. The story itself could affect our ability to reach a solution—which in and of itself is useful and relevant to human understanding. *Reverse the Odds* uses the story of the Odds and their need for a reconstructed world as part of the impetus for responding to questions about cancer cells.

4. *Rule using*: These problems typically involve constraints and rules and require people to "find the most relevant information in the least amount of time."[18] Games, in general, embrace this type of problem solving, as games involve some type of constraints and rules of play. These rules can help to guide players to identify, classify, or collect the appropriate types of data. In *Happy Moths*, for instance, a player receives rules about how to evaluate a moth, and then she can apply it to each of the images she views and use it to effectively categorize them.

5. *Decision making*: Decision-making problems involve picking one option from among a set of options; or selecting from among choices that have consequences. These types of problems can be simple or complex. In *Who Is the Most Famous*, players need to make snap decisions about the most famous person with a specific first name, and

this has consequences for one's score and leaderboard position, as well as the collective data on celebrities. In *Apetopia*, the player needs to decide which color gate is most similar to the background color.

6. *Troubleshooting*: These problems involve some type of diagnosis of a problem and debugging. This might involve continual hypotheses, testing, and retesting to figure out what is wrong. In the tutorial segment of *EyeWire*, players click on different areas of a neuron to "color" them in to learn more about the process of mapping the brain. The system provides feedback to the player, such as by labeling correctly mapped areas the color green and incorrectly mapped areas with the color red.

7. *Diagnosis solution*: Diagnosis solution problems relate to determining the most appropriate course of action or identification of a condition, such as the diagnosis of medical issues or an insect problem, or a course of action for a student who is struggling with a topic. *Cell Slider* participants evaluate potentially cancerous cells, helping scientists to get better at diagnosing cancer in future patients. Similarly, in *Biogames*, players help to properly diagnose malaria in cells.

8. *Strategic performance*: Strategic performance problems involve executing a series of tactics to serve a strategy, typically in real time, whereas there may be multiple solutions and implementations, depending on the context and content of the decision. Examples could be a teacher deciding what type of curriculum to use and how to use it for a particular learning situation or an obstacle-course contestant figuring out the best solution to approaching the course. Many knowledge games could incorporate this approach, such as *Play to Cure*, where there may multiple ways to solve a given problem of finding the best route through space and players need to figure out the best course of action. *Phylo* is similar to *EteRNA* in that players seek the best way to sequence genetic material, and both games involve elements of strategic planning and algorithm creation. *Phylo*, for instance, uses colored blocks to represent genetic components, and participants move these blocks from side to side to make the best match with the target sequence.

9. *Case analysis*: Case analysis problems are often used for learning and teaching purposes; they simulate reality and typically involve consideration of the context. Law students may work through real-world legal cases and business students may think through a specific marketing case to help them more authentically practice the skills necessary in their fields. *The SUDAN Game*, as it has been described, since it has not

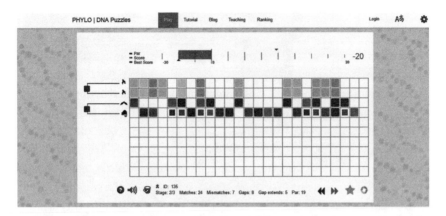

In *Phylo*, players manipulate and match sequences of colored boxes to help trace genetic diseases. Source: Jérôme Waldispühl

been released yet publicly, could be seen as a case problem, in that it simulates the many variables associated with Sudan's divided tribes and enables players to iterate through possible solutions. Game worlds in general could be seen as cases with elements that simulate aspects of reality, such as how the *Civilization* series illustrates portions of running a civilization and managing resources over time, or how *Fable III* expresses the consequences of a key environmental choice on the fictional world of Albion.

10. *Design*: These problems involve understanding the best way to design a product, process, object, or system and can be applied to designing anything, from this book to a school system to a healthcare policy. Design problems within games could include *EteRNA*, where people figure out the best way to devise a new RNA molecule based on a given set of constraints. Games themselves are also design solutions, as discussed further in chapter 2.

11. *Dilemmas*: Dilemmas are unpredictable and less structured problems. With dilemmas, there is no clear answer, and there are pros and cons and benefits and drawbacks that must be weighed to reach a solution. Knowledge games may be a good way to think through dilemmas; for example, how to handle a child who is disruptive in school or how to talk to a veteran who might be suffering from PTSD, particularly as there is no clear answer and there may be many diverse views that are better served by inviting as many perspectives as possible. In *School-Life*, for example, participants are given different scenarios related to

Players in *SchoolLife* interact with digital characters to explore ways people may handle bullying situations. Source: Jeff Orkin, CTO / Co-founder GiantOtter

bullying and need to respond to them. The goal is to use this information to help better understand and then simulate the complexities of bullying and people's different reactions to it, and then use this to alter the game and eventually enhance people's empathy and consequently, reduce bullying.

There are other ways to categorize problems. Newell and Simon, for example, describe three types of problems, which as we can see overlap quite a bit.

1. *Social/interpersonal problems*: These are problems that focus on the relationships among us, such as dating, parent-child relationships, or racism.
2. *Natural/ecological problems*: These are scientific types of problems, such as choosing an antibiotic for a particular illness or figuring out how much water a plant species needs.
3. *Technological problems*: These are related to how to use and create new tools, such as creating sustainable materials or a better toy for language learning. This includes four subtypes: inventions, designs, troubleshooting, and procedures.[19]

We can also look at how structured problems are, how abstract they are, or whether they have already been solved. For instance, a problem can vary on a spectrum from being highly structured to ill structured. Ill-structured problems are typically ones that are not yet resolved, so people could postu-late multiple perspectives and solutions for them, as in the case of *EteRNA*'s RNA design or *Foldit*'s protein structure challenges. There may even be more than one possible answer, and people could explore many different solutions to discover the optimal one, as in *The SUDAN Game*'s possible paths to peace or *SchoolLife*'s potential ways to handle bullying.[20]

Logical problems are usually more structured and constrained, while di-lemmas and design problems are typically considered the most ill struc-tured because they may not yet have a solution or any correct response.[21] Problems could even span several problem types, such as in the example of *Foldit*, which has elements of both algorithmic and design problems (using the Jonassen framework), or in the case of purifying water, which could be a technological, social or scientific problem in Newell and Simon's scheme. Individual steps to a given problem could even contain different types of problems. Complex problems, such as how to implement a fairer healthcare system, how to handle global warming, or what to do about a particular mental illness, could have a variety of subproblems within them, which could be labeled as dilemmas, design, diagnosis, decision making or trou-bleshooting problems (using the Jonassen framework), or all three of the Newell and Simon problem types.

Introne, Laubacher, Olson, and Malone apply the concept of the "wicked problem," which was first named by Rittel and Webber. Wicked problems could be considered ill-structured, large-scale, systemic, complex dilemmas, or "problems for which no single computational formulation of the prob-lem is sufficient, for which different stakeholders do not even agree on what the problem really is, and for which there are no right or wrong answers, only answers that are better or worse from different points of view." Schoder et al. label most social and political problems as wicked problems, including those "related to environment, healthcare, social welfare, education, and security," such as the healthcare, global warming, and mental illness prob-lems listed above. Wicked problems could be figuring out what to do about children who are struggling in school, the high rates of recidivism for prison-ers, the contradictory issue of high unemployment and lack of skilled work-ers, or how to cure illnesses that have both social and scientific causes, such as heart disease.[22]

Potentially, ill-structured, complex, and "wicked" problems are a good fit for knowledge games. After all, if the solution were so clear, we would not need so many people with unique perspectives and experiences trying out and exploring different solutions. However, any of the types of problems Jonassen describes could be a good fit for a knowledge game, depending on the needs, goals, and audience. Even a problem that has many current, already-established solutions, for example, may still need more people to resolve or reconsider it, to offer new perspectives or solutions, or to find new ways to represent it. We may know how to identify cancer cells in images, but we still need lots of people to help do it. We may have a current solution for how to handle students who are failing their classes (such as kicking them out of college), but there may be other interventions or preventative measures we can take. And likewise, an ill-structured problem may be too unwieldy, complicated, or open-ended for a game. Games are not necessarily the best way to solve a given problem. Addressing an immediate climate crisis, such as a coming tsunami, blizzard, or hurricane may be better achieved through immediate, emergency evacuations than through people staying in harm's way and using a game to find and mimic alternate ways of surviving the perfect storm. (Although using a game for planning and preparations could be quite useful and appropriate.)

Dilemmas and wicked problems may be the more obvious types we should consider solving through games, but we should not forget the other types of worthy problems that games can potentially support. Ultimately, we cannot just state that we want to solve problems through games. We need to explore the types of problems best solved through different types of games, and this is still an open question.

Appendix A lists examples of some current knowledge games (particularly those described in this book) and the primary problems they seek to explore. While these examples can serve as a guide for the types of problems that can be solved through games, we should not let it confine us. We should be experimenting with the types of problems better solved with games and the types of game elements that can better support their solutions, with the caveat that this is difficult due to the inability to fully, accurately, and dynamically represent and contextualize problems in games.[23] In fact, uncovering the mechanisms and processes behind any problem solving is difficult—whether in a game or not.

Games and Problem Solving

Before we start to criticize or laud the problem-solving ability of games and game players, we need to understand more about why games might be particularly adept at supporting problem solving. Yes, we know that they set up problems for players to find a solution to through gameplay. We understand that games are even games, in part, *because* they present problems and a space in which to solve them.[24] But what are the specific elements, characteristics, and aspects of games that may make them especially amenable to problem solving? How are games distinct from other media, or even crowdsourcing platforms, that they might provide a different dimension to problem solving? In this section, I break down the aspects of games that may make them effective for problem solving.

Human players can work in tandem with a computer or computer-supported experience, such as a digital game, which can potentially optimize the problem-solving advantages of each. There are problems that are either more easily solved by a person or by a computer, and there are problems that are optimally solved by both in concert with each other. Human beings, for instance, are great at pattern recognition and can even discover complex patterns where none intentionally exist.[25] It reminds me of when, at an old hotel in New York City, my then-18-month-old daughter viewed the seemingly random pattern of moldings on the lobby wall, and declared them as being filled with owls. People are always looking to make a meaningful structure out of chaos. Based on the discovery of a possible pattern, human beings will make a hypothesis about what was found, explore their environment, and then continue to readdress the pattern until they are satisfied with their assessment of it. Humans are also better than computers at other tasks, from completing surveys to labeling and tagging images, to writing up reviews, to fixing audio-transcription mistakes, as suggested by the typical tasks listed on Amazon's Mechanical Turk platform.

Computers, on the other hand, are great at solving well-defined, formalized problems, such as arithmetic queries, very quickly.[26] However, when it comes to wicked problems, ill-structured problems, dilemmas, problems that are poorly defined, or with aspects of tacit or implicit knowledge that are not clearly able to be codified, computers fall short.

Working together in tandem, however, human beings and computers can be effective solution-creating systems. This relationship harks back to the concept of human computation, introduced in chapter 1, whereby computers can take the algorithms created or generated by human manipulation and process them faster, such as in the case of *Foldit* or *EteRNA*.[27] Computers can also provide choices (such as what to do about a bully) and consequences (such as how a bully might react) and pose them to people to see how different players react or respond to subsequent iterations. This is different from books or written versions of a problem, which cannot possibly provide a complex system based on one's unique combination of choices. Computation, as Shaffer argues, "makes it possible to develop simulations that dynamically enact and reenact parts of the way we understand our world." Likewise, in my research on environmental sustainability and games, I explain how playing within a game world such as *Fable III* helped players imagine and think through the consequences and implications of their behaviors and decisions and consider ones even beyond the effects simulated in the game.[28]

Schoder et al. identify the following as being particularly useful ways that information systems or computers can better support complex problem solving in tandem with people: (1) enhancing the deliberation of possible solutions to a problem; (2) helping people navigate their social interactions and networks to assess strengths of relationships; (3) individualizing or customizing information based on a person's preferences; (4) highlighting or suggesting more relevant techniques and just-in-time hints to people; and (5) handling, managing, reconstituting, and regurgitating large amounts of data for further manipulation.[29]

Players can work with other players to share perspectives, contribute, and/or come to synergistic conclusions. Problem solving often occurs in teams because the process of solving frequently requires abilities and expertise beyond any individual's capabilities.[30] We can point to concepts such as distributed cognition and collective cognition, terms discussed at length in chapter 5. Some problems require many eyes, hands, and minds working in concert.

Collaborative problem solving may even lead to more optimal problem-solving outcomes. Collaborative problem solving typically involves a shared understanding, appropriately working on a problem together, and interpersonal organization. For instance, people need to work together directly to figure out, negotiate, and execute tasks to solve a particular problem. Yigal Rosen looked at collaboration among human teammates, as well as with computer or digital artificially intelligent nonplaying characters. To do this, Rosen observed a small sample of 14-year-old participants working on a zoo-related task with either a human or computer partner. The results of the tasks suggested that in both conditions, participants created meanings and outcomes that they would not have been able to do alone. Even working with a computer character helped participants enhance their collaboration skills, particularly because they had fewer opportunities to be in disagreement in the human-to-human interaction. (The computer was programmed to purposely misinterpret or disagree about the particulars of the problem and its tasks.) Having points of contention among people and experiencing alternate perspectives help support collaborative problem solving and, potentially, better solutions, so long as the differences and disagreements are not insurmountable.[31] This is similar to having a town council meeting where everyone is from the same political party—if they all share similar viewpoints, they do not need to respond to each other's arguments and, ostensibly, grow in their own perspectives.

Rosen's study suggests that a game that includes collaboration with a digital avatar or in-game guide may be beneficial to bringing out the best and enhancing the problem-solving skills of the player, even if the players do not (or do not need to) collaborate with other players. However, in Rosen's study the problem posed was neither highly complex, ill structured, nor contextualized within a game, and it was not required to be completed by more than one person. Moreover, the inclusion of other people could help in solving problems (even if a player is not working directly with them). Trying on others' points of view, taking on different identities, or exploring other roles can help people see problems from new perspectives, in addition to encouraging the use of new tools, frames, language, terms, and norms, all of which may lead to better problem-solving outcomes.[32]

Games may also, at least theoretically, open up problem solving more to the public, possibly along with the data, frameworks, contexts, and results related to the problem further enhancing its problem-solving potential. Lakhani et al. argue that scientific problem solving should be more open, transparent, and public-facing. In practice, however, much scientific research and problem solving is closed, constrained, and secretive, which means that we do not leverage the experience and expertise of the broader community. The currently "closed" nature of science is due to several factors, such as incentives for publications, grants, individual achievement, and securing proprietary information for competitive purposes, which in turn lead to tenure, promotions, and the like.[33]

Lakhani et al. looked at InnoCentive (see chapter 1) and found that when problem information and the solving process is opened up further and shared with the public, it can be very effective, particular because it means that there is a diversity of opinion, interest, expertise, and experiences working on that more open problem, rather than before. Overall, their results suggest that having diverse problem solvers was advantageous, as was supporting more open access, transparency on the processes, and collaboration in problem solving.[34] More about the connection between social interaction and problem solving can be found in chapter 5.

Games, motivation, and problem solving are interconnected. Games do not automatically motivate problem solving simply by the fact that they are games; however, interrelationships do emerge. Eseryel et al. investigated the interconnections among motivation, problem solving, and engagement in games, particularly in games focused on learning. They empirically studied motivation by looking at a problem scenario in *McLarin's Adventure*, an educational MMOG geared toward eighth and ninth graders, and their findings suggested that enhanced engagement in the game and what they call prior problem representation was related to overall problem representation in the game and resulted in better learning outcomes. Providing appropriately timed and balanced challenges and problems throughout the game experience seemed to relate to more engagement and motivation in the game.[35]

Conversely, motivation and engagement can influence problem-solving behavior in a game. Research by Ryan, Rigby, and Przybylski suggests that maintaining motivation through a game will encourage players to continue to complete tasks and overcome obstacles, spurring them to spend more time solving given problems. When people are more motivated, they will also put more effort into solving a particularly difficult problem. Likewise, in InnoCentive, time spent on a problem was correlated to devising more effective solutions, further suggesting that having more time to practice might help make perfect.[36] More about motivation is explored in chapter 4.

Games can provide a story, context, or analogy to aid in problem solving. Providing a story context for a problem can help in a number of ways. Stories can offer people new perspectives or reframe the problem so people do not get stuck applying less-useful problem-solving methods. A story can also help people avoid getting caught up using numbers or abstractions that are less personally meaningful. This is similar to how Jonassen describes story-based problems, which typically embed a more solvable or quantifiable problem within a story context. Shum et al. describe how story can help us relate to data and numbers and provide meaningful context. They explain that creating a story, such as characters, plot, worlds, and interactions around facts and raw information, helps to contextualize a problem and illustrate its human dimensions rather than just providing bare numbers.[37] For example, simply inputting the numbers or cancer data in *Play to Cure* might be less effective than providing a meaningful game context in which people can play with those data sets (in this case, a spaceship adventure). Likewise, in the "crowd verification" game *StormBound*, the designers "hide the math" so that instead of seeing numbers, players interact with magical streams of energy and rune symbols.[38]

Similarly, in the Facebook-based game *Fraxinus*, players need to compare and create leaflike colored patterns. The patterns represent genetic data on ash trees and the Chalara fungus, and the game is being used to check "draft" sequenced genomes to figure out the correct way of assembling the sequenced strands of data, and to investigate how this fungus is able to infect ash trees at the genetic level.[39] The abstraction of genetic data into leaflike objects makes the information easier to grasp quickly and to manipulate in order to solve a complex problem. In *The Restaurant Game* or *SchoolLife*, the context of a restaurant or bullying scenario is more meaningful and motivating than having people just submit to a database what they would do in a restaurant or in a bullying interaction. Moreover, in the case of *SchoolLife*, scenarios and stories in the game could even evolve based on players' actions and their consequences.

A story could also be integrated into an ill-structured or not typically story-based problem to help highlight its complexity, draw people in, and motivate further engagement, or even to illustrate alternate approaches to the problem, such as in the case of *Forgotten Island*, which uses the story of a mysterious island and robot to engage players.[40] There are even areas unique to the human brain related to analogical or story-based problem solving. Ultimately, story is also what may help a problem's solution become accepted and established more widely in the public. The context and story around the knowledge gives it meaning and shows us how we can use it and integrate it into our lives.[41]

Games can encourage alternate problem-solving paths. Games can help people move beyond what is called functional fixedness, or being fixated on the typical function, convention, or solution of a given object, relationship, or problem.[42] A game can help shift "stuck" perspectives so that new ideas, solutions, or ways of thinking can emerge. Games can also help players reflect on, probe, and challenge current models of the world and keep us from being fixated on already-in-use or problematic solutions.

Games can use story for this purpose, but there are other strategies. For instance, games can reinforce or reward alternate ways of using familiar objects or concepts, show or simulate familiar tools in new settings, or provide novel contexts around a given problem. They can enable and even encourage seeking multiple ways to solve a problem by rewarding several routes to success, enabling players to personalize their own paths, make their own choices, or try out alternative avenues based on a player's interests, abilities, style, preconceptions, or experiences. This may also help players realize that there is more than one way to solve a problem and that other perspectives are valuable and essential, freeing them to consider unconventional routes. Even if a person has not experienced those other paths, just knowing that they exist can be powerful in avoiding being stuck in one's own perspective.[43]

Even if a game's rules and paths to success are clearly defined, such as in *Cell Slider* or *Reverse the Odds*, players can personalize the game to their desires. For example, people can choose which goals they want to pursue and when, such as when they want to earn new potions by evaluating cancer tissue images in *Reverse the Odds* or wait the two minutes for them to replenish.[44]

James Gee explains how games spur players to go through a cycle of questioning and probing the world, where they make a hypothesis based on what is experienced and learned, test that hypothesis and reconsider the world in light of the effect of trying it out, and then glean feedback from those interactions, possibly building a new hypothesis about the world. He calls this the "probe, hypothesize, reprobe, rethink cycle," and it is integral to problem solving and may be useful in helping players overcome misconceptions. In other words, a player cannot just formulate a solution; she needs to test the solution in regard to the given problem, see what the outcome is, and decide whether that solution is appropriate. This is different from stopping after the first solution, not waiting to see what the feedback looks like, and just moving on, possibly leaving behind a fallacy or erroneous solution in one's wake. By presenting problems using real-world simulations and dynamic systems, games can authentically and meaningfully provide opportunities for players to actively explore and test out hypothesized solutions and, through in-game feedback, experience consequences and decide whether hypotheses are valid.[45]

Another key way that games help us avoid canned solutions and replicating misconceptions is through puzzlement.[46] Puzzlement occurs when a world does not live up to a player's expectations (such as when a cat in a game world suddenly sprouts wings). Puzzlement piques players' attention and drives them, potentially, to try to understand the discrepancy between their expectations and what is happening in the game, thereby trying to "solve the problem" of this incongruity.

Games can authentically situate and simulate a problem within a dynamic system, though this has its limits. Many games can dynamically simulate and situate real-world problems, experiences, concepts, or interactions by serving as miniature models of the world.[47] This is akin to Papert's concept of the microworld, where people can push and pull on a game's boundaries and constraints, explore properties, and experience elements that could not be felt in the real world.[48] Every action that a player takes in a game relates to direct or indirect (implicit or explicit) hypotheses about that world. Players can test out assumptions and see if they hold true. This iterative activity builds players' understanding of the part of the world being simulated and also can potentially help them explore related problems.

Squire points to two types of models that scientists use, both of which could be expressed through a game: idea models, which express concepts (such as the water cycle or centrifugal force), and predictive models, which can be used to predict future possibilities or outcomes based on variables and inputs (such as what would happen if two trains were moving in a particular direction at a certain speed). While designers may provide the general constraints and factors into the model simulated through a game, the play of the game may reveal emergent possibilities or relationships that were otherwise unable to be seen. The game's system provides a framework for the problem, which players can embody, leading hopefully to novel interconnections and unexpected outcomes.[49]

Games can also model and embody complex decisions and outcomes and even promote a player's own mental simulations, enabling the player to more easily visualize future possibilities.[50] In a study on ethics and games, I found that those players who were playing a game version of an ethical scenario (in *Fable III*) were more able to conceptualize, and to incorporate into their thinking, the future consequences and possibilities than those who just worked through the same ethical scenario that was written out on paper.[51] This type of thinking is often called "systems thinking" or an understanding of the dynamic interdependence of systems.[52]

When the tools, information, and contexts used in a game are authentic, it also more likely facilitates real-world, transferable problem solving. Shaffer describes an example of students learning about how a farm works by doing the work of a farm. By solving real problems—whether how to milk a cow or how to tap maple syrup—they were more able to imagine what working on a farm was like. These learners inhabited the epistemic tools and frames of being a farmer and were immersed in farm culture, which helped them solve the typical real-world problems that farmers need to solve (such as how to milk a cow). Likewise, Shaffer argues that the most effective kinds of learning games enable participants to work on real-world problems, such as running a Wild West trade route in *Frontier*, managing an amusement park business in *RollerCoaster Tycoon*, or being a limnologist and trying to save a lake from being polluted in *Citizen Science*. But the real-world problems posed in *Frontier* and *RollerCoaster Tycoon* do not necessarily have a direct, outside-of-the game referent. In knowledge games, people are working on real problems and this work has an actual effect on the out-of-game world. Players may use the relevant elements, variables, and resources that are currently present in our world to help solve these problems. Thus, games can represent not just the materials, tools, and information about a real-world matter, but they can attempt to represent the real-world issue itself.[53]

Games may help novices gain the necessary expertise to participate in problem solving. One of the most common difficulties with general problem solving is that, particularly with collaborative or cooperative problem solving, different people arrive at a problem with different abilities, experiences, skills, approaches, misconceptions, and biases. Good games, however, enable people with all different types of backgrounds and expertise levels to become skilled enough to participate in the problem solving in the game.[54] Games enable access to the necessary expertise so that the problems presented in a game can potentially be solved by more players.

In a well-designed game, novices are not just thrown in; rather, their learning is scaffolded and supported through practice problems, tutorial rounds, beginner levels, trial and error with feedback, didactic instruction, peer-to-peer support, the game's mechanics, and/or a clear articulation of the boundaries of the game's system. Games can level content, enable the practice of necessary skills, and test players to ensure they have appropriate levels of expertise and skills before moving on to a more complex problem.[55] Players can then continue to build their skills, abilities, and knowledge to gain the necessary mastery to approach the game's open problems.

Moreover, knowledge and problem solving can be stored in the objects, characters, and environments of the game, as well as in the minds of fellow players, freeing the player to explore the world and make novel connections. Games can also break up larger problems into more palatable sub-tasks and mini-problems.[56] For example, *EyeWire* does not just throw you into the main task but provides a lengthy tutorial to test your assumptions and help you build enough skills and expertise to apply it to real-world problems. In both *EteRNA* and *EyeWire* a chat window appears within the game interface where players can query others for help and advice in real time, which may be particularly helpful for novices as they start to understand the lay of the problems and even more for entrenched players when they are ready to ask more detailed questions.

Getting novices up to speed effectively is important in knowledge games for at least two reasons. First, it opens up the problem to more problem solvers, which can lead to more diverse perspectives, abilities, and opinions. Second, subject experts may not always be the best at solving a problem. Nonexperts may have perspectives and connections to offer that experts become blind to once they learn enough about a field. For example, when Lakhani et al. looked at InnoCentive, they found that sometimes greater problem-solving success and efficacy were seen when people worked on problems in fields adjacent to, or even completely outside of, their core expertise. Their research even suggested that the farther a person is from the problem's field and expertise, the more likely they can solve the given problem. This is counterintuitive, but it could mean that novice participants were able to move beyond functional fixedness or see new perspectives on old problems that experts could not see.[57]

On the other hand, Eseryel, Ifenthaler, and Ge argue that experts have more linkages among concepts and a more integrated cognitive structure, which may aid them in the application of their knowledge toward solving problems and being innovative. Likewise, Bransford, Brown, and Cocking examined the function of expertise, and their analysis suggested that the experts' abilities to solve problems depended more on knowledge organization and structure than on a deeper understanding or facility with a topic.[58]

These claims have not yet been empirically tested with knowledge games specifically, so they should be studied further. However, these findings suggest that inviting nonexperts or amateurs from diverse fields to solve problems could contribute to more effective solutions and outcomes than keeping a problem closed and away from the public.

Limits to Problem Representation and Manipulation in Games

In the previous section, I described characteristics of games that might make them more amenable to effective problem solving. Games, however, are never perfect, unbiased models of a problem or problem space. Any time you pose a problem, such as a research problem, you necessarily include some and leave out other information. You may choose a particular language to describe and frame the problem or select specific images or videos to use to display the problem. You set the agenda for the research and elect the questions to include in a survey; you decide which population of people to sample; and you may choose to focus on particular factors, such as gender and age, rather than socioeconomic status, when creating a model of the problem.

Likewise, one book may discuss quantum mechanics very differently than another, or a particular documentary may frame genocide differently than a given newspaper article. Similarly, designers frame problems in games. They use certain tasks, mechanics, and goals to construct the game's play while downplaying other variables. Games, and their designers, shape the world of the game based on what matters to them, what they value, and what types of factors and questions are essential to them. Biases (not necessarily negative ones) are inherent to design; the predispositions affecting a game's design could be culturally, socially, or economically based, or they could be the result of knowledge gaps, ignorance, or blindness to other ways of thinking. Biases could even be from factual errors, incorrect data, or even structural issues, such as sexism or racism.[59]

Thus, the game, and the way it is designed, may enable or constrict certain paths toward a solution, reinforcing or influencing which solutions rise to the top. Two games may pose the problem of how to analyze cancer data very differently, which then affects how any problems are represented and solved. *Play to Cure* transforms cancer data into a spaceship path, based on a need for finding irregularities in the data. *Reverse the Odds* asks participants analyze and categorize images of potential cancer cells, which in turn enables them to earn potions that help in solving mini-puzzles. And *Cell Slider*, which uses similar cell-categorizing tasks as *Reverse the Odds*, provides no narrative or mini-puzzles but simply asks people to view and evaluate images (in fact, the *Cell Slider* creators worked with the *Reverse the Odds* team to provide the virtual cell-categorizing mechanism). Another team could then take that same set of data or images and cell-categorizing technology and use it quite uniquely in a game experience. That team might ask different questions, offer up a different narrative around it, or invite people to analyze different aspects of it.

Just like with any research project, scholarly book, documentary, or mediated representation of a problem, how something is designed has an effect on the types of solutions devised, how those solutions are communicated, and how they become established as knowledge. A player's ability to solve a given problem through a game depends, in part, on how the problem is designed, how that design is communicated, and how any biases are managed. When designers are transparent about their choices, reflect on their preconceptions, and communicate their approaches, it is more likely to help others appropriately evaluate any solutions from the game and decide what types of conclusions can be drawn. However, it is often difficult to reflect on one's own biases since they are part of who we are, they are not typically intentional, and they may not even emerge until a completely different paradigm or cultural context arises.

Kurt Squire argues that there may be value in considering not just the game's design and its rules, as well as the underlying biases and assumptions of the designers, but also to explore and reflect on the emergent interactions that occur due to those rules and constraints. Simply reading through a game's rules or a design document is not enough, and not even necessarily relevant. Rather, playing or inviting others to play out those rules and experience what emerges may be more useful in evaluating how a game's design will affect the solutions procured.[60]

In the next chapter, I explicate the interconnections between motivation and knowledge game playing by looking at research from volunteering, crowdsourcing, and citizen science, as well as from games (both educational and mainstream).

CHAPTER FOUR

Motivation

Ever wanted to map neurons in the brain? Even if this hasn't been your lifelong dream, the Seung Lab makes it a feasible fantasy with a game called *EyeWire*.[1] The premise is that anyone who can color within the lines can also map a neuron (brain cell). The impetus for *EyeWire* is that a computer cannot construct 3-D representations of neurons from 2-D splices of cells, so researchers need human help to figure out where a neuron's twists and turns lie.

When you first begin *EyeWire*, you see a 2-D cross-representation of a neuron on the right side of the screen, and a 3-D representation of the same neuron on the left. To "play" the game, you zoom in, scroll, click, and color in the 2-D neuron cell pieces. The places where you color are then used to build a 3-D neuron representation.[2] The game continues like this—clicking, scrolling, and coloring in a neuron—for 30 to 40 more minutes until you are finally done with the initial training. At that point, however, you still have a few more hours of practice brain cells to complete before you can start to color "open" neurons—or ones that do not have 3-D representations generated for them yet. It also becomes clear that the game does not veer from those original three mechanics (click, scroll, color). In other words, after hours of playing *EyeWire*, the game still consists of clicking on, coloring in, and scrolling through 2-D cell images to make 3-D cell representations—and nothing more. It also means that it could be a few hours of skill building before you are actually contributing to science, which is not a lot of time in the long run but may be a big investment for the casual contributor.

Despite this, *EyeWire* is one of the most successful citizen science or knowledge gaming projects to date. Its creators boast more than 160,000 players from 145 different countries, and they have published hundreds of papers based on the tasks performed in the game.[3]

So why would anyone want to play *EyeWire*? To motivate their players, the *EyeWire* team has designed a leaderboard, weekly competitions, feedback on each neuron you "solve," and regular e-mails to registered players giving hints on the game and encouraging more play. People may even inherently love the idea of spending hours clicking on neuron cells and may

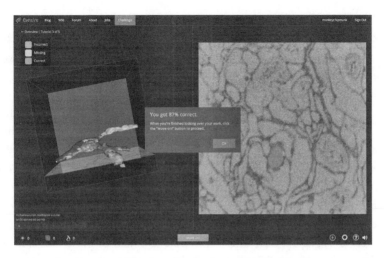

In *EyeWire*, players color in brain cells to create 3-D models of neurons. Source: EyeWire

gain a feeling of satisfaction at having contributed to science. After all, some players spend 40 hours a week playing. *EyeWire*, and its players have contributed to our knowledge of neurons, such as the starburst amacrine cell's structure, which is implicated in how eyes detect motion.[4] But is this a fun, engaging, or motivating game—and does it even matter?

Why Play Knowledge Games?

Encouraging people to play and continue playing is integral to the success of any game, especially knowledge games. After all, you cannot collect, analyze, or process enough data, solve enough problems, or gather enough perspectives unless you have enough players. *EyeWire* cannot map the brain with its small research team; instead, it relies on many players to do the dirty work to complete the neuron mappings and contribute to our knowledge of the brain.

In this chapter, I question why people might play knowledge games. We know that, in general, a well-designed interactive mix of goals, feedback, actions, and rewards can be motivating in a game.[5] But what are the nuances? Are there factors that are unique to knowledge games? By understanding what compels potential players, knowledge game makers and researchers can better adapt current games or design new games, whether by adjusting language around the game, adding or subtracting communication features, or shifting mechanics to be more in line with overall project goals—or completely reimagining knowledge game paradigms.

What does recent research tell us about the factors that motivate game players? What does it suggest about the motivation to help, learn, collaborate, volunteer, and contribute? How can we benefit from this research when designing, developing, and implementing future knowledge games?

Many mainstream games, such as the *Grand Theft Auto, Call of Duty, Mass Effect,* and the *Sims* series are highly engrossing endeavors that have been designed specifically to motivate gameplay. Additionally, these best-selling commercial off-the-shelf (COTS) games may be multimillion-dollar, big-budget efforts, with large development teams who can afford top-notch storytelling, visuals, sound, characters, and voice acting, all of which can contribute to making a game as engaging as possible for its chosen audience. Some games, such as the free-to-play *Farmville* or *Candy Crush* series, are designed to engage players and to generate as much profit as possible. Designers may use data generated from their game players, such as when people make in-game purchases, to optimize profit from current and future games.[6] They may also incorporate principles from social psychology and motivation research to further motivate play and profit.

Knowledge games, by contrast, are often made with low or no budgets and small teams, and by researchers rather than seasoned game designers, artists, programmers, and talent. Few knowledge games can feasibly have expertly crafted open story worlds such as that of *Dragon Age: Inquisition,* exceptional fight mechanics such as that of *Batman: Arkham City,* or exquisite landscapes such as that of *The Elder Scrolls: Skyrim*—nor should they necessarily have these elements. For better or worse, knowledge games are typically molded with the primary intention of gathering accurate information (*Cell Slider*), enabling interactions with appropriate models (*Foldit*), assessing human behavior (*The Restaurant Game*), or gathering opinions (*Who's the Most Famous*), and not solely to motivate gameplay for the sake of playing the game. Some knowledge games may have slicker packaging than others, such as *Reverse the Odds* or *Play to Cure*, but the primary design goal is to produce accurate, appropriate, or useful knowledge.

When someone has the choice to play the latest *Grand Theft Auto*, the top-selling console game of 2013 with an almost billion-dollar budget,[7] versus lower-budget *EyeWire*, or *Clash of Clans* and *Candy Crush Soda*, two of the currently most popular social mobile games, versus *Reverse the Odds*, most are not going to choose to play the knowledge games, even if similar motivating elements are present in each. There's just no comparison.

Though we do not yet have any empirical evidence on this, I can initially hypothesize that people are not necessarily playing knowledge games because they are the most fun and enjoyable games out there (though these aspects could still be important) but because they also offer other types of experiences, such as the ability to help, contribute to society, and to change the future of knowledge. Thus, we cannot simply discover how *FIFA*, *League of Legends*, or *World of Warcraft* motivate players and apply it to knowledge games because the audiences, rationales, and motivational elements could be vastly different. Comparing players of knowledge games to those of COTS games is like grouping together people who volunteer at a zoo with those people who go to the zoo just to visit the animals and saying, well, they are both at the zoo and both like animals so we should analyze the two groups and their motivations similarly.

So what insights can we highlight right now about motivation, particularly as there are few knowledge games available and little research on what motivates people to play them? While we wait for more of this research to be conducted, we can look at studies on citizen science, crowdsourcing, and volunteering more generally, as well as research on games in general, including educational games.

Why are participants playing knowledge games? What really drives someone to seek out and participate in these types of experiences? Why would people provide free labor, volunteer time, effort, resources, and game playing particularly when, as we see from *EyeWire*, knowledge games might be tedious, time-consuming, low-budget, and even downright boring?

What Is Motivation?

Defining and characterizing motivation is tricky. Ryan and Deci explain that motivation is simply about moving someone to do something, such as playing a game or volunteering for an organization. Motivation is the driving or underlying force to explain why a person wants to play, plays, or continues playing a game, whether *EyeWire*, *Grand Theft Auto*, *Farmville*, or *BioShock*, though we cannot always apply the motivations for one game to another. More formally, in relation to citizen science participation, Raddick et al. explain that motivation is "a mental construct that a volunteer uses, consciously or unconsciously, to explain their behavior, arising out of a combination of the person's mental state and properties of the situation they are in."[8]

Ryan and Deci explain that there can be different levels (quantity) of motivation, as well as different orientations or types of motivation (quality). Their self-determination theory explores two motivation types: intrinsic and extrinsic motivation. In intrinsic motivation, people are "doing something because it is inherently interesting or enjoyable," which comes from within their own self and inner interests. In extrinsic motivation, people are "doing something because it leads to a separable outcome," where motivation stems more from external factors, forces, symbols, and objects, such as money, badges, social status, or another external reward. Typically, the popular understanding is that intrinsic motivation leads to better, more innovative, and creative outcomes, but extrinsic motivation is not always weaker, as it depends on other factors. In fact, it is difficult to conceive of any activity, no matter how beloved, which does not also involve external social, cultural, or economic pressures and influences, in addition to inner desires. People can internalize external forces and factors such that they can become more integrated into one's self and more self-determined. In general, Ryan and Deci's research found that those activities that supported people in their need to feel connected to others and helped people feel more competent and have agency and autonomy over their choices and behaviors were also more likely to buttress new ideas, skills practice, and knowledge acquisition.[9]

In this chapter, I first investigate specific activities related to knowledge gaming, such as volunteering and citizen science, to see how motivation plays out.

Why Do People Volunteer?

According to the U.S. Bureau of Labor Statistics, about a quarter of all Americans volunteered in 2014; this rate is about the same as 2013.[10] Should we consider participating in knowledge games, crowdsourcing, and citizen science projects as volunteer work? Some crowdsourcing projects, and perhaps some citizen science projects, have financial incentives associated with them, such as Threadless and InnoCentive, so an analogy between these types of activities and volunteering is not perfect. However, can we draw parallels between unpaid or volunteer work and knowledge games to provide insight into some possible motivational factors?

In a classic 1998 study on volunteer work, Clary et al. investigated why people participate in unpaid work or volunteering. To do this, they surveyed 467 people and asked them to rate the importance of different reasons as to why they volunteer. Using this information, the researchers identified six functions of volunteering, including people's values (e.g., because it is important to help humanity), understanding (learning through experience), enhancement (helping someone grow and feel good about himself), career (helping someone build her resume or provide access to a paid job), social (helping people connect with others and meet new people), and protective (helping reduce negativity and escape reality).[11]

Shye critiqued the Clary study and other previous studies on volunteering. He argues that Clary et al. were too focused on bottom-up empirical research and did not first start with vetted motivational theories. Shye questions whether some responses of participants in the study, such as "service to others" (which is definitional to volunteering) and "a friend's suggestion" (which is possibly a stage to volunteering or circumstance) are actually even motives. To Shye, a motive is why someone wants to do something—not the resources that enable someone to volunteer or the circumstances that prompt that effort.[12]

Thus, instead of simply asking people why they volunteer, Shye asked participants to identify the benefits of volunteering from among 16 options that he established from his own and previous research. He looked at the needs and functions that volunteering filled, rather than how one got to be a volunteer, and found that the ability to make friends and feel a sense of community were the most important needs met by volunteering. Shye also surveyed nonvolunteers and found that they "do not seem to appreciate as much as volunteers, the fact that volunteering enables one to express one's beliefs, or one's unique personality . . . these kinds of information can be valuable for volunteer recruitment and retention."[13]

Likewise, Wilson finds that social context is important to volunteering. Being a part of a social network helps people hear about opportunities to volunteer or meet other volunteers, "but also to believe that other people will share their volunteer impulse, that they will not be left in the lurch by others shirking their responsibilities." Different types of social connections can affect how and whether a person participates and can be variably important depending on the stage of volunteering.[14] For example, what you need, socially, to propel you to start to volunteering may be different from the types of social connections you need to further engage you in continuing to volunteer after you begin.

Moreover, there is research that prosocial activities in general help make people happier—and that contributing makes you feel good and feel needed. Some populations may experience fewer depression symptoms and improve overall mental health through volunteering.[15] Morrow-Howell, Hong, and Tang interviewed and surveyed over 400 older adults and found that most participants felt that they were "better off" from volunteering (50%) or "a great deal better off" due to their volunteer activity (31%), with only 18% saying there were no benefits or limitations to volunteering, and only one person claiming to be worse off. Sixty percent explained that their families were benefiting from their volunteering activities as well.[16]

More broadly, people are motivated to work toward a common social goal. Batson et al. created a model of motivation specifically related to contributing to societal issues and needs, which has four specific types of motivation. The first is egoism, which is a desire to increase your own well-being; second is altruism, or wanting to increase well-being of another group at your own expense (of time or resources); third is collectivism, or wanting to enhance the collective well-being of your own group; and finally, the fourth is principalism, or wanting to imbue qualities that matter to you.[17] Although these models have not been directly applied to knowledge games, in the future we may want to apply Shye's, Clary's, or Batson's frameworks to test the motivations for playing these types of games.

Why Do People Crowdsource?

Another activity similar to knowledge gaming, though again not a perfect parallel, is crowdsourcing. As discussed in chapter 1, crowdsourcing involves contributing some type of activity, analysis, or data. Not surprisingly, it's particularly important for crowdsourcing to have a crowd, and the concept relies on many people providing information, knowledge, and effort. Thus, understanding what makes crowdsourcers tick is of utmost important to those designing platforms for it and those researchers and corporations using it or integrating it into a knowledge game.

First, let's look at InnoCentive, an online community for solving pressing scientific problems. The purpose of InnoCentive is to enable scientists to openly and transparently work on many complex problems in the hope that leveraging the expertise of many people can help provide solutions. This type of platform seeks to challenge the common lack of sharing of knowledge among scientists, which may impede the ability to create new knowledge.[18]

Lakhani et al. tried to parse out the reasons why people participate in InnoCentive. They surveyed more than 300 InnoCentive participants, including 68% of people who successfully solved a problem and 34% who did not, during a set period of time. People who solved the problems spent about twice as long on the problems than nonsolvers, with an average of around 74 hours of total time for the solvers, and 35.7 hours for the nonsolvers. The top motivators of participation for solvers in InnoCentive were winning award money, enjoyment of the types of problems, the intellectual challenge of solving problems, having free time, and a desire to enhance skills. Intrinsic motivations, such as enjoying the problems, were typically stronger motivators, whereas some extrinsic motivations, such as wanting to enhance one's career or earn greater social prestige were found to be negatively correlated with being a winning solver.[19]

Daren Brabham critiques the Lakhani InnoCentive study, noting that most of the participants on this platform hold PhDs and that the financial incentive on InnoCentive was potentially quite high, which may have affected their results. Brabham conducted his own studies on motivation and looked at two other crowdsourcing platforms: iStockphoto and Threadless. In his study "Moving the Crowd at iStockphoto," he surveyed more than 600 participants in iStockphoto, a website where participants can buy and sell original photographs in an online marketplace that uses microtransactions for each image sold. Brabham's survey results suggest that the primary motivators of participation on the site from the sellers were "the desire to make money, develop individual skills, and to have fun." Many iStockphoto participants felt that their participation on the website helps hone their photography and media-creation skills and is a creative outlet. In other words, beyond the money earned, the iStockphoto members also took pleasure in the very act of creation and in the development of skills to aid that creation. The participants in Brabham's study skewed male, white, married, middle or upper class, with an average age of mid-30s, and typically a bachelor's degree or higher.[20] This is a similar demographic to the participants on Galaxy Zoo, for example, though a bit younger.

The differences in demographics and incentives between iStockphoto and InnoCentive may have resulted in some of the differences in primary motivation between the two studies, suggesting that the context and constituency of participation can correlate to different motivators. However, in reviewing both studies, there is still quite a bit of overlap in the primary motivators of the platform participants.

Brabham also studied Threadless, an online community where creators submit artwork for T-shirts and other apparel, which the community then votes on to decide which shirts actually get made and sold on the site. He interviewed 17 participants and found that the four primary motivators were earning money, developing creative skills, expanding one's potential to do future freelance design work, and following one's interest in the Threadless community. A fifth motivator that emerged was what participants called an "addiction" to the Threadless community, in that they felt constantly connected to the community—wanting to check results, post blog entries, or discover new T-shirt designs. While Brabham notes that the source of the addiction may be the technology, I argue that the technology facilitates social feedback and approval, as well as creative outlets, which may be gratified through the platform. These participants are enmeshed in the Threadless community and feel that their continued participation matters and is integral to the overall success of the endeavor. Likewise, the Threadless creators ensure that its contributors feel as though they are stakeholders in the process; they listen to the community and improve it through their user's suggestions.[21] While Brabham's study's small sample size limits the generalizability of the results, it offers insight into possible motivators for participants in a successful and engaged community of contributors. It also suggests that community is sometimes a strong motivational factor, but it is not always a top factor, depending on the type, context, and interactions of the community. The elements that determine when and under what conditions a community is a primary motivational factor should be considered further.

Studies of other crowdsourcing projects reveal other motivations such as one's inner passion for a project, interest in being recognized for one's contributions, desire for personalized feedback from experts, and a drive for advocacy. However, one of the major issues with applying motivational factors gleaned from crowdsourcing projects to other projects is that study results are sometimes contradictory and context of a project matters a lot.[22] Depending on the type of project (e.g., business versus science), motivations might vary. Zhao and Zhu argue that few studies incorporate motivation theories or include meta-analysis to determine the extent to which contexts of the crowdsourcing activity, extrinsic-versus-intrinsic motivators, or other characteristics affect motivation.[23]

Why Do People Engage in Citizen Science?

What motivates citizen scientists? When reviewing projects such as Galaxy Zoo or eBird, Rotman et al. explain that "few [citizen science] projects have reached the full collaborative potential of scientists and volunteers." Volunteers are often considered peripheral to a citizen science project, even though their participation drives its success. Citizen science efforts by definition need to motivate citizen (amateur) volunteers to participate in a project to fulfill its goals, such as collecting enough data on bird locations or bug identifications. Yet attempts to recruit and maintain citizen scientists have been uneven, and not every project is successful.[24] Although currently little empirical research exists on this topic, I will describe three studies on citizen science motivation: one on Biotracker, one on Galaxy Zoo, and one on a suite of citizen science games.

Rotman et al. looked at Biotracker, a conservation and energy-focused project, and surveyed 142 respondents and interviewed 11 people to understand why they participated in it. The researchers explain that in general, two aspects of volunteer participation are affected by motivational factors: "the initial decision to participate in a project and the ensuing decision to continue this engagement once the initial task is completed." They found that the primary reasons stated for initially joining Biotracker were an interest in the project, such as curiosity about the content of the project, previous hobbies or activities related to the tasks, or a desire to build a resume for a related career.[25] For example, participants felt initially that their involvement in the Biotracker project would help them gain specific skills, such as plant identification. Over time, however, other aspects of the project engaged them further, such as community involvement and advocacy or understanding the related environmental issues adequately to share and communicate them to others.

Factors affecting the scientist and citizen scientist interactions also emerged in the study. Rotman et al. found that the nonscientist volunteers felt intimidated by the scientists and their expertise level, and that there was a lack of trust between the two groups. They also found that most scientists did not understand or were not aware of the actual motivations of their project's citizen participants, which is something to consider more deeply as we make more knowledge games.[26] Both of these are challenges for knowledge game makers. How do we enhance trust between players and researchers? How do we truly understand the motivations of a game's players so that we can better design the experience for them as well as enable more effective knowledge production?

Bowser et al. also looked at Biotracker but with "Millennial"-generation students who were not already participating in the project. They found that the motivational elements for this population are game elements (such as earning badges or competition), social aspects (membership in a community), and personal drives (wanting to learn or have fun), but not an interest in wanting to contribute to scientific knowledge, which is often seen in typical volunteers to citizen science, as the Rotman et al. study indicates. While the Bowser et al. study results could be extrapolated and used to suggest which elements could attract new players to participate in knowledge games, many of the interviewed students stated that they were not likely to use Biotracker in the future (only 14% were likely or very likely to do so), despite the presence of motivational factors.[27] Thus, how do we attract and retain this population of players—and should we bother?

Raddick et al. looked at a sample of Galaxy Zoo participants to try to tease out their motivations. Using open-ended interviews, they first identified 12 motivations, such as "I want to contribute to original scientific research," "I am happy to help," and "I am interested in science." Next, they surveyed a larger sample using these motivations as a guide and inviting other motivations as well. Even with a low response rate, around 11,000 Galaxy Zoo participants responded to the survey. The researchers found that the top reasons for participating were: wanting to contribute to science, an interest in astronomy, a desire to discover something new, and the beauty of the images. Almost 40% of those surveyed were most motivated by wanting to contribute to scientific knowledge.[28]

What about citizen science projects that are also games or have gamelike aspects? In a study with a much smaller sample size than the Galaxy Zoo study, Iacovides et al. interviewed four volunteers who play *Foldit* and four who play *EyeWire*. They questioned participants on both their initial motivation and which elements sustain their engagement with the two games. The researchers found that the participants were initially interested in the projects because of an intrinsic interest in science—not because it was a game they could play. However, the researchers observed that although the gamelike aspects did not pique interest initially, game elements may help to prolong participant engagement. In particular, they noted that two elements (which do not necessarily need to be applied to a game)—"a sense of progression" and "team play"—could be useful for extending engagement if meaningfully and appropriately applied to the project's goals and content.[29] In sum, related studies suggest that the fact that a knowledge game is a game may not be what is particularly attractive or meaningful for the player at first, but a game's elements and mechanics (such as goals and progression toward them) and other characteristics (such as the communication tools and community around it) may serve to further engage the audience and sustain their interest. This also means that if you are implementing a knowledge game, you may not be looking for typical game players at first but instead people interested in your project's topic or its implications.

Iacovides et al. offer three recommendations for those considering designing these types of games to sustain engagement:

1. Use teams, with a mix of collaboration and competition among players. Allow for both single-player and team-player modes.
2. Tie game mechanics directly to the goals of the project. Any badges or points should be related to player contributions and knowledge in the project and not be separated from the project's content or goals.
3. Employ communication platforms such as forums or chat functionality so that participants can collaborate, ask questions, and discuss strategies or contributions.[30]

Considering the small sample size, however, all results and recommendations should be considered preliminary.

Prestopnik and Crowston investigated a suite of games, *Citizen Sort*, which includes *Happy Moths*, *Forgotten Island*, *Living Links*, and *Hunt and Gather*. Their analysis indicates that a platform that encourages people already excited about a game's topic may also inadvertently discourage those not interested in it, and vice versa. How is this possible? Prestopnik and Crowston suggest that if people want to be immersed in their hobby, or one aspect of it, a knowledge game may not fulfill them as much as engaging in that hobby more directly. For example, in *Forgotten Island*, players need to collect and classify items, such as photos of moths, as well as explore locations and engage in the story; however an entomology hobbyist who just wants to classify the items may feel that the game or story elements are extraneous or frustrating.[31] At the same time, the scientific or collection elements in the game may not attract or sustain game players who are interested primarily in the story and game elements and do not want to spend time on classification activities.

Prestopnik and Crowston's research suggests that some games, particularly those that try to include elements to attract both avid game players and science buffs, may actively discourage both. Consider *Reverse the Odds*, which has social and puzzle elements that are separate from the game's cancer cell image-identification tasks. Some players may just want to classify the cancer images and not bother with the Odds story or puzzle solving.

Others may simply want to focus on the puzzle solving and not have to deal with doing the image-identification exercises. Therefore, both constituencies may end up less satisfied with the game than if it just focused on one goal (entertainment or identification). *Reverse the Odds* is a new game, and no research currently exists on player motivations, so whether this plays out remains to be seen.

Prestopnik and Crowston's study generated three design approaches, which we can apply to knowledge games.[32]

1. The first design approach involves what Prestopnik and Crowston call "gamifying" knowledge-producing tasks: incorporating scores, leaderboards, competition among players, or other game elements. In *Happy Moths*, for example, players receive scores based on their completion of moth-identification tasks.

2. The second approach makes the main knowledge-producing task a minor part of the game. This approach wraps game elements around tasks, such that there are in-game justifications for those tasks, even if the game itself is not fully integrated with or relevant to the project's goals. This is the most challenging approach of the three. An example is *Forgotten Island*, where players explore a mysterious island and engage with a story about a robot while also performing specimen-identification tasks as part of the gameplay.[33]

3. The third approach is one where the knowledge-producing task's completion provides direct entrance into the rest of the game, or hastens or unlocks the ability to play it. This seems to be close to *Reverse the Odds*, in that you need the potions earned from the cancer cell image-identification tasks to help you complete the game's puzzles.

As a fourth approach, designers should seek to go farther than just gamifying tasks within a fantasy world, using tasks as entrance into the "fun" part of the game, or wrapping the game around tasks, and instead try to create a game that fully integrates the goals of the project with the gameplay and goals, themes, and flavors of the game. With this approach, the game elements would not be just tacked on top of the game; rather, they would be fully partnered and holistically experienced.

Finally, we should also ask what motivates scientists to invite citizens to participate in their research. Why would scientists want to collaborate with everyday citizens? Rotman et al. found a couple of main factors. First, scientists want to gain access to lots of data collection or analysis, such as making observations, classifying specimens, or collecting photos. However, they also want to remain the arbiters of the data, the ones who add the value while the citizens do the "dirty work." Second, scientists want to contribute to public education and teach others about their field through citizen participation, and they want to help the public better understand policies around science and the role of science in humanity.[34] Scientists' and game players' motivations do not always align, however, which is useful to keep in mind when designing and using knowledge games.

Why Do People Play Games?

Now that we have considered motivations for volunteer work, crowdsourcing, citizen science, and citizen science games, it is useful to fold in research about motivations for game playing more broadly. Yet, as mentioned earlier in the chapter, we cannot just take what makes games generally motivating and apply it to knowledge games and expect to make better, more captivating, or more effective games, just like we cannot always apply techniques that work with COTS or mainstream games to educational games.[35] However, knowledge games are still games after all. While for many knowledge game players, motivation to volunteer and contribute to society are most salient, if we want to increase and enhance knowledge gaming's reach and design more relevant and relatable games, we need to also consider how these factors intersect with game elements to further motivate and continue to engage players.

Before we venture farther, I need to briefly clarify the ever-expanding role of games in the general motivation of a variety of activities that take place in the real world. For instance, in *Gamify*, Brian Burke explains how using game mechanics and experience design can help motivate people to achieve their personal goals. Likewise, Jane McGonigal expands on the call by psychologist Mihaly Csikszentmihalyi for real life and real work to be more like games, particularly to avoid boredom, anxiety, and alienation. For example, we might take a dull health issue, such as taking a pill every day, and make a game that can motivate people to pay more attention to (and even enjoy) the pill-taking routine. Paul Darvasi turned his high school English class unit on *One Flew Over the Cuckoo's Nest* into an alternate reality game (ARG), where he invited the students to "join the asylum." He gave "game" points for everything from journal entries to videos related to the book's content, which equated to the students' grades. He used the ARG to motivate his students to learn humanities-related content and themes of the book and also to further encourage the practice of media literacy and writing skills. McGonigal argues that such "gamified" experiences may push people to work harder or deeper because transforming life, work, or books into "a game" makes that which is not fun, fun, and even may help us fix our broken reality.[36]

I am not arguing for or against the use of games or game elements to motivate activities outside of the game—sometimes it is appropriate and sometimes it is not. But I want to reiterate that knowledge games are not about gamifying the world outside games to make things more fun, game-like, or more motivating. It's not about placing game mechanics on research methodologies. Knowledge games are, themselves, games. They may motivate change outside the game, but they themselves are contained as games, and you can contribute solely as a game player. In other words, you can play a knowledge game and have a personally gratifying experience, or you can play the game to help spur change that is outside of the game, or both—but the knowledge that is produced is conducted within the game. On the other hand, the increasing blurring of boundaries between games and the real world may make this distinction moot. Where does a game stop and the rest of the world begin?

In this section, I summarize research findings and principles related to games and motivations. While this area of research is still understudied, it is useful to consider how we can apply any findings to knowledge games.

Games should support the different needs, play styles, and motivations of players. Games should enable different flavors of play and satisfy different needs—after all, players may be motivated by varying kinds of gameplay. Richard Bartle theorizes that there are four player types: "Killers wish to act on (i.e., kill) players; Socializers wish to interact with players; Achievers wish to act on (i.e., achieve within) the virtual world; and Explorers wish to interact with (i.e., explore and manipulate) the virtual world."[37] What's your type?

Nick Yee expands on Bartle's framework using empirical research to map player types. He analyzed survey data from thousands of online game players (such as those who play *World of Warcraft*, *EverQuest*, or *Star Wars Galaxies*). He found there are three main categories or clusters of motivations: achievement, social interaction, and immersion. The achievement cluster relates to ways that a player can gain and maintain power or reputation in the game. The social-interaction cluster centers on how people relate and form relationships with others in the game. The immersion cluster focuses on how people can become part of the story or world of the game, escape the real world, and take on new roles.[38] Players can have different aspects of these clusters matter at different times, and the clusters can overlap with each other to create a unique window into each person's motivation for a continued game-play experience.

In addition, the personal characteristics of players also matter. Researchers have cited differences in gender, race, culture, reading ability, and personality as potentially related to responding to and being motivated by diverse types of gameplay experiences.[39]

While this research focuses on entertainment-centered games, rather than knowledge-producing experiences, how might it apply to knowledge game players or would-be players? What would be the analogous knowledge game player types?

Games can help people tap into their intrinsic interests and desires. Games may fulfill our unique passions, goals, and interests, which relates to the concept of intrinsic motivation, introduced earlier in this chapter. For example, Crowley and Jacobs researched kids and found that experiences, such as games, which fit their interests and specific "islands of expertise," such as cars, animals, moviemaking, or space, drew them in more and motivated them to more deeply practice associated skills.[40] Malone et al. created a taxonomy of intrinsic motivation that may be useful for understanding knowledge games. Its three initial categories—challenge, fantasy, and curiosity—were later expanded to also include cooperation, recognition, and competition. Daniel Pink's taxonomy, also pertinent, describes the intrinsic needs for autonomy, mastery, and a sense of purpose, which some games can provide.[41]

Ryan, Rigby, and Pryzbylski applied self-determination theory to video games and created the player experience of need satisfaction (PENS) measure to determine how people consider the play experience in terms of five dimensions: competence, autonomy, presence/immersion, intuitive controls, and relatedness. Nick Yee argues that while there are oftentimes specific motivational reasons for playing a game, sometimes we just play because it taps into that which makes us human and part of humanity. The concept of flow also relates to intrinsic motivation and interests. Mihaly Csikszentmihalyi has argued that particular activities and experiences, such as play and games, can trigger a feeling of flow, or "the satisfying, exhilarating feeling of creative accomplishment and heightened functioning." Jane Mc-Gonigal explains that this intense engagement with the world makes us feel "fully alive, full of potential and purpose . . . completely activated as human beings."[42] What actually elicits flow may be different types of experiences for different types of people, but when we are in a flow state we are so highly engaged with that flow-prompting experience that we will continue with it.

Providing choices in games has been found to increase motivation and perceived control. Having just the right amount or type of choices can help people feel autonomy, agency, and a sense of control in games. For example, giving players appropriately leveled choices and challenges may help people feel higher self-efficacy and feelings of empowerment in a game. Likewise, Salen and Zimmerman argue that choice is fundamental to games and that the system's meaning arises from the choices a person makes in a game.[43]

Choices, even just the customization of characters or environment—such as changing the appearance of one's avatar, color, and clothing—matter to a player. Having choices may even be related to enhanced learning. In a study by Cordova and Lepper, kids who were able to have choices in their game were more engaged in the game, more able to handle greater challenges in it, and their overall actual learning and perceived competence increased.[44] This research suggests that everything from minor (changing one's avatar's name) to major (enabling choices to affect the game storyline or environment significantly) could be related to higher motivation. Too many choices, however, can be overwhelming and stifling, so we also need to find the right balance of choices.

People like to solve problems and overcome challenges in games. Many people want to master and "win" at something to feel like they are competent and able, particularly those people who are achievement focused, as described earlier. In a game, achievement and mastery can mean anything from overcoming an obstacle (such as successfully defeating a Deathclaw in *Fallout 3* or jumping over a barrel in *Donkey Kong*), to accomplishing a large, multifaceted task (such as solving the structure of an unknown protein in *Foldit*).

Goals need to be appropriately difficult, but at least theoretically surmountable. In Malone's work on intrinsic motivation, education, and games, he suggests that challenge is integral to motivation, which includes cultivating players' feelings of self-efficacy and competence. People feel motivated to build new skills, solve novel problems, and perform well particularly if any challenges are well matched to their abilities. Challenge in a game can be created through a variety of means, such as altering or progressing difficulty levels of problems, providing multiple or flexible goals, obscuring information, or randomizing tasks. However, if a game is too easy or too difficult or follows the wrong progression of challenges, it will decrease a player's motivation.[45]

Eseryel et al. looked at the MMOG *McLarin's Adventure*, and their findings suggest that it is important to sustain players' motivation throughout the game by continuing to provide new, more challenging problems while also making them feel like they eventually have mastery and success with the challenges they are given. Just playing the game is not necessarily going to lead to further problem solving; rather, players need to be engaged in the game's tasks and feel like they are progressing in their proficiency. Eseryel et al. offer three types of interaction that should be designed to sustain motivation, including interface interactivity, narrative interactivity, and social interactivity. These need to work together to sustain motivation, they argue.[46]

People want to connect with other people, whether directly or indirectly, and participate in the community or affinity space around or within a game. Not all knowledge games provide a social or collaborative component, nor do they need to. While a game's development and design is typically a collaborative effort, the actual gameplay, problem solving, or puzzle pondering does not need to be done in conjunction with others, depending on the needs of the game and the project's goals, or the computational and communication platforms used by the game.

However, for some types of game players, and in some contexts, solving problems with others (or even in competition with others) may be particularly motivating. Youth often play games with friends, either in the same room or virtually. In online games, many people are motivated to play because they are interested in socializing, developing relationships, working with others to solve problems or complete tasks, or co-designing stories and role playing together.[47]

Likewise, building a large network around a knowledge game may not be as important as finding ways to have players encourage others to play. Having experts ready to chat or offer advice to newcomers may help people who are just starting out. Having experienced group members reach out to players and give tips or hints might also be useful for motivating new players initially and for continuing to motivate more expert players.

People are curious. People are naturally interested in the unknown and want their curiosity piqued in a variety of ways. Likewise, game players like surprises and mysteries and want to play so they know what is going to happen next. My toddler daughter loves to watch "egg videos," where people open plastic eggs with toys in them, because she loves the surprise of finding out what is inside.

In Malone's classic study, he suggests that curiosity (how much is aroused and satiated in a game) is implicated in motivation.[48] One of the factors in arousing curiosity is an incongruity of expectation or lack of consistency, which leads to questions about why, how, and what is happening, further motivating the person to find out the answers. (See chapter 3 for more on puzzlement.) If we expect that a dog cannot bounce on a pogo stick, and it suddenly bounces on one in a game, that surprise can draw us in deeper.

People want to contribute and help. As seen with volunteering, crowdsourcing, and citizen science research, people may play knowledge games, and games more generally, because they want to help out others and contribute to the common good. Social context, coupled with the opportunity to volunteer, can draw players in even more. We have seen how social context can be one of the most motivating factors in initiating and continuing volunteer work and how a person's social network plays an important role in helping him find out about volunteer opportunities and nudging him to attend them.

In relation to games, people may spend hours and significant effort contributing to a game's wiki, informally or formally tutoring each other, or responding to questions about a game on forums. For example, in some MMOGs, newcomers to the virtual world can ask questions to more seasoned veterans to learn basic commands, hints, or tips. In *EteRNA*, new participants are able to chat directly with experts to ask questions and get oriented to the gameplay. People may also organically form groups, such as guilds in MMOGs that can gather and share resources and go on missions together. Players may even form problem-solving posses where they work together on problems such as those in *Foldit* or InnoCentive. To understand these efforts better, we can look at Batson's model or the volunteering theories and research described earlier in the chapter, though these particular frameworks have not yet been applied to knowledge games.

Prestopnik and Souid's research provides some insight more directly into knowledge games. They argue that although players in theory want to help out and contribute when they start playing a game, such contribution can feel like work if other elements of engagement, such as narrative, are missing.[49] For example, their *Forgotten Island* places the tedious knowledge-producing tasks in the background of a story-intensive game so those tasks do not feel like too much work. Thus, knowledge game designers may want to frame their game as a way of giving back to the community and as a way of exercising citizenship in a global community of learners and players.

People want to experience a story, narrative and/or emotional experience. Not all games have an explicit story or narrative, but those that do can motivate people through their desire to experience, complete, or co-create a story or narrative experience. Most of the gameplay in *Gone Home* involves interacting with objects and items that help to tell the story of a family living in the 1990s. The player is motivated to interact with these objects to uncover the storyline. Similarly, in *Dear Esther*, the game is a poetic journey through deserted beaches and caves, which tells pieces of the avatar's life story as you travel through the game environment. The uncovering of this story as you travel is the core of the gameplay.

Related to this, Malone discusses fantasy as a key motivator. He uses Walt Disney World as an example and describes how children imagine and experience new worlds, see themselves in a new role, or fulfill wishes through play and games.[50] Likewise, my toddler daughter enjoys playfully pretending to be a different character each day—sometimes a dinosaur, snake, cat, or mouse—and involving her family in the fantasy.

Bringing the Principles Together

How all these principles specifically contribute to sustaining motivation in knowledge games is still unknown and understudied. Specific design elements in games may support self-efficacy and a feeling of competence, which work in tandem with other aspects of a game, such as social interactions within the game or the cultural context around it, to motivate players. Motivating players may encourage further problem solving in a game. Understanding, dissecting, and assessing these specific elements, and the conditions under which games help motivate and sustain engagement, is difficult because of the complexity and interrelationships among them.

In the next chapter, I explore the social dimensions of knowledge games. While many knowledge games are effective in part because of social interaction among players, not all of these games need to incorporate social interaction, such as cooperation or collaboration. To understand the possibilities and drawbacks, I explicate the primary research and frameworks that relate to social interactions in crowdsourcing, citizen science, and games and how they may be relevant to knowledge gaming.

CHAPTER FIVE

Social Interaction

Who is the most famous Leonardo? Do you immediately think of DiCaprio? Or do you think of da Vinci?

Who Is the Most Famous? asks just this question and compares your answers with other players'. In each round of the game, you get just a first name and are asked to type in who you think is the most famous person with that name. For instance, if you're presented with "Fred," you then have just 15 seconds to make your best guess as to the most famous Fred's last name. The last time I played, Fred Flintstone was winning with 33% of the votes, while only 7% said Fred Astaire. You are awarded points based on the speediness of your answer, whether you had the most popular answer, and whether you had any of the top answers that other people typed in. You get a score and rank after each round. Thus, to earn more points, you need to align with the "social ether" and choose the answers that you think are more popular while also competing with others to see who will end up at the top of the leaderboard.[1]

While the purpose of the game may be to crowdsource famous people's names and to test your recall of and familiarity with luminaries, *Who Is the Most Famous?* actually rewards you for successfully "gaming" the likely demographic's general knowledge of celebrities.[2] The game values the top answers, which means that if the purpose is to find a wide range of views or attract a diverse population, it is falling short by purposely rewarding the most popular or most mainstream views to persist. However, if its purpose is to try to measure broad perceptions (and any changes in those perceptions over time) or to guess a particular demographic's perspective on celebrities, then it's an effective game. The goals of the game designers, however, are unclear, as there is no information on the game's landing page beyond just the game itself.

The game is also repetitive—the core mechanic is typing in famous people's last names, and this does not change from round to round. The game does get more interesting, though, particularly when it queries classically "non-Caucasian American" names, such as Carlos, or female names, such as Kelly. Who ends up being most popular for Carlos or Kelly? How does this reflect cultural perceptions or structural inequalities? For Kelly,

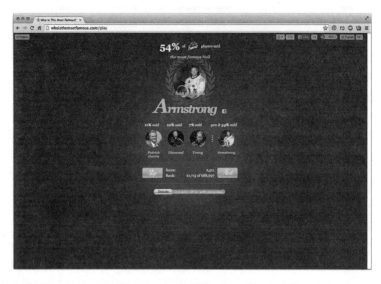

Who Is the Most Famous? participants quickly enter the most famous person with a particular first name. Source: Yoni Alter @yonialter

the most popular name might be on the music and acting side (such as Kelly Clarkson, the "Kelly" winner when I played), rather than from science, literature, or sports, which may reflect how difficult it is for minorities or women to enter and attain superstar status in certain fields, or at least to keep their real name in them.

At a broader level, it would be interesting to see how many athletes versus music versus movie stars versus scientists versus authors versus fictional characters are mentioned, segment this data set by gender, race, or other demographic markers, and investigate it longitudinally—but there does not seem a way for the public to access this data set. What happens with gender-neutral names, for example?

Even in just my brief play-through there were some surprise results that I would have liked to investigate further. "Mark" turned out to have "Mark Twain" as the most "famous," with 24% of the players entering his name (particularly interesting as that was his alias).[3] Isaac came up with "Isaac Newton" as the most popular, whereas Leonardo had DiCaprio was listed as more popular than Leonardo da Vinci, who earned second place. I am sure these historical figures would be pleased to know that even among all the celebrities of today, they still come up at or near the top.

As I played *Who Is the Most Famous?* I wondered, do other players affect your gameplay? Is this a "social" game if you are not directly collaborating or communicating with others? Although other people do not directly interact with you or your famous person guesses, the perceptions and perspectives of the social ether are continually affecting your gameplay. Fame itself is inherently social, as you are only famous if you are known to people. You are both competing and collaborating with other players—trying to ensure that you align with others and have the top answer while hoping that your total points are higher than everyone else's. Do social interactions affect knowledge produced in a game?

The People Principle

Under what conditions should people work together to solve problems and contribute data and perspectives through games? How do communities around or within games affect the game experiences? How does this affect how knowledge is produced? This chapter delves more deeply into the social and collaborative aspects of knowledge games and how social elements might be leveraged to further encourage participation, enhance problem solving, and help us learn more about ourselves and our lives.

As we saw in the last chapter, people may be initially motivated to volunteer, crowdsource, or participate in citizen science because of having an interest in science or wanting to contribute to the common good. However, they may be engaged to stay longer and participate more enthusiastically due to, for example, social interactions or the desire for social approval. Could we apply this to knowledge games? Can these games strengthen and build on social interactions in ways that deepen participation, motivation, and problem solving?

Several researchers have noted the importance of social interactions to playing games, and how games can in turn help reinforce social bonds. Games might even encourage collaboration in some cases, and they can potentially support collaboration between human beings and computers, rather than just among people directly.[4] Some knowledge games may benefit from effective collaboration, social interaction, cooperation, or competition, whereas some would not. This chapter delves into the types of problems, tasks, play, experiences, and situations that do or do not benefit from social interactions.

Yet games are often considered quite antisocial. For instance, a common perception is that they cause aggression or reclusive behaviors such as those embodied in the popularized image of a gamer locking herself in her bedroom or basement for hours or days, without food, sleep, or even bath-room breaks (such as with *Hikikomori*).[5] However, the purpose of this chapter is not to uncover whether games cause people to be more or less antisocial or sociable, or whether they replace or transform social interactions. Rather, I consider the conditions under which collaboration, cooperation, and other types of social interaction may or may not support the goals and needs of knowledge games.

Collaboration versus Cooperation

Many terms and processes that evoke social interactions have been applied to games, problem solving, and knowledge making. We have crowdsourcing and crowd wisdom, distributed and collective cognition, collective intelligence and cooperation, coordination and competition, social computing, human computation, and human-computer interaction, to name just a few. Before we go too far I want to clarify and distinguish among some of the key terms.

First, what is the difference between cooperation and collaboration? Cooperation typically involves the mutually beneficial sharing of goals, resources, and responsibilities that could not as effectively or successfully be accomplished with one individual, organization, or institution. Collaboration, by contrast, involves more direct sharing and interactivity among people or groups. It represents "the process of shared creation: two or more individuals with complementary skills interacting to create a shared understanding that none had previously possessed or could have come to on their own."[6] In other words, with collaboration, two or more people or organizations come together to solve something that is more than either could have accomplished alone.[7] Roscelle and Teasley explain that collaboration is a process of negotiation and sharing, where participants communicate and deliberate meanings and coordinate activities in order to create a shared understanding of a problem and solution.[8]

However, Hall critiques Roscelle and Teasley's definition of collaboration, explaining that it implies that collaboration needs to happen by parties working together simultaneously, and that it is only focused on problem solving. According to Hall, this definition does not take into consideration how communicative technologies, such as games, can support collaboration across distance and over time. Hall instead prefers Mäkitalo-Siegl's definition that collaboration is when "participants pursue joint goals and solve shared problems and build mutual understanding of some particular issue . . . an active process where learners enter into a joint activity and adopt common goals that bring them together to perform tasks or solve problems."[9] In sum, cooperation involves players contributing individually, but toward a common goal, and collaboration involves more direct interactions to problem solve together. Based on this distinction between cooperation and collaboration, *Happy Moths*, *Cell Slider*, and *Cropland Capture* would be primarily considered cooperative games because they coordinate players to submit data or perspectives, which are then aggregated and reconfigured to become more actionable information. With cooperation, each player's goals are the same, but they are not interacting with each other directly to accomplish the game's tasks. In cooperative games, individual data are contributed, and the collective sum of data is equal to the total sum of individual activities. In collaboration, however, individuals interact in dynamic ways so that the total sum of their collective activities is *more* than the sum of what could be created from its individual players.

It is unclear whether *Who Is the Most Famous?* is a strictly cooperative game. When you play the game, you are not directly working with or communicating your guesses to other players. However, as the game progresses and you start to discover the popularity of various names, you may alter your answers to better fit your prediction of which names "will win." Therefore, you end up experiencing a sort of virtual "dynamic conversation" with other players. Your answers influence others' answers, and others' answers implicitly or explicitly affect your own answers in the game. You are learning from others synergistically even if you do not directly interact with anyone.

Likewise, in Galaxy Zoo, a citizen science platform, when categorizing galaxies, you are not working together directly with other participants; however, there are peers and experts in the forums and available through a chat function who can answer questions and work through problems with you, making aspects of Galaxy Zoo both cooperative and collaborative (see more about Galaxy Zoo in chapter 6). Most knowledge games and crowdsourcing projects lean more heavily toward the cooperative side of the spectrum, though *Foldit* and InnoCentive can be considered collaborative platforms because they can involve people working together directly to find a solution that is more than the sum of the individual's abilities (although, you do not have to work with others to participate in those platforms). In this chapter, I focus on the aspects of collaboration and cooperation in knowledge games.

"If You Can't Do It Alone, Work Together"

Can we take collaboration and cooperation to the extreme by adding lots and lots of people? Are there benefits to large-scale collaborative and cooperative processes, particularly when we integrate this with Web-enabled platforms, such as online knowledge games? Can we solve complex problems with more people working together in dynamic arrangements across distance and time? "Collective intelligence," "distributed cognition," and "collective cognition" are all terms being used to describe slightly different aspects of computationally supported, people-involved problem solving (although their usage overlaps quite a bit).

"Collective intelligence" was first coined by Pierre Lévy and suggests that knowledge is shared among the collective society, rather than only in an individual's mind. Collective intelligence is a concept about the emergent properties of social interactions, such that when people work together they can better problem solve and make decisions. For example, people have looked at collective intelligence's role in stock market prediction, social tagging or bookmarking (such as via Delicious), guild activities in *World of Warcraft*, the participation of fans in the co-creation of the *Trainz* experience, or in spoiling *Survivor*.[10] However, uses and definitions of the term "collective intelligence" widely differ. There are over 500 papers on collective intelligence alone, yet no common definition exists. In their book *Wikinomics*, Don Tapscott and Anthony D. Williams define collective intelligence as a type of mass collaboration with particular application in business and which relies on four principles: openness (sharing ideas and resources), peering (peer modification), acting globally, and sharing (relinquishing control). The Center for Collective Intelligence at the Massachusetts Institute of Technology takes a more interdisciplinary approach to collective intelligence and uses this concept to consider questions from cognitive science (such as how the brain is organized and how this relates to how people are organized in groups), computer science (such as how to create human-machine hybrids), and business (what strategies can we use to employ large-scale people problem solving most productively), as well as other disciplines.[11]

A related term, "distributed cognition," refers to individual members of a spread-out or distributed group of people who are processing information and tasks, similar to the earlier definition of cooperation. In distributed cognition, resources such as time or abilities may be distributed among the group and completed concurrently to complete a massive task or solve a complex problem.[12]

Collective cognition, however, differs slightly from distributed cognition. In collective cognition, people interact, share, and compare so that the collective experience is more than merely a sum total of individual activities. This enhancement derives in part from emergent dynamics and interactions that take place when people work and process together. Citing Kozlowski and Ilgen, Hung explains that a new level of cognitive processing emerges in which "these distributed cognitive processing and actions of the team members are interindividually integrated into the team's cognitive processing and behaviors as if the team is one functional cognitive entity/system." Collective cognition emerges when people work together in a holistic way, so that the resulting new group processes, which could not be predicted by the individual behaviors of the people in the group, arise dynamically.[13] This is similar to the definition of collaboration explored earlier.

Most of these terms, such as "collective intelligence," have been applied to group dynamics more recently, but are these concepts of distributed or collective brain power really so new? Hollan, Hutchins, and Kirsh argue that regardless of whether we are problem solving alone (in a game or not), we are never doing it completely in a vacuum or separate from human activity. Button argues that "distributed cognition" is simply another name for cognitive science as a whole and that using the term to describe social phenomena just extends the idea that cognition is not done in a vacuum from other people. Likewise, Gee contends that all thinking, reasoning, and knowledge are distributed such that different people know different things, and different people, tools, and platforms help us to think, store, and use knowledge; the "really important knowledge is in the network—that is, in the people, their texts, tools, and technologies and, crucially, the ways in which they are interconnected—not in any one 'node' (person, text, tool, or technology), but in the network as a whole."[14] In other words, the interconnections in and around a knowledge game, rather than any one aspect of the game (player, designer, or game platform), are what really matter, and knowledge is always and has always been distributed across people and objects.

Distributed cognition, collective cognition, and collective intelligence are also related to concepts such as "human computation" and the "wisdom of the crowd." Human computation focuses on how problem solving power can be distributed not just among people, but among people in tandem with computers. This is similar to Gee's aforementioned notion that people, tools, and technologies are all a part of a system of distributed knowledge.[15]

The term "wisdom of the crowds" refers to James Surowiecki's book and concept of almost the same name; it suggests that a crowd's collective expertise is usually more accurate than any one individual's expertise. This may seem counterintuitive because we would assume that someone who is an expert in the field should be better at making decisions than any large-enough group of people would be at coming up with a collective decision. Surowiecki, however, gives examples of when the crowd's wisdom prevailed, such as when he asked audiences to estimate the amount of marbles in a jar, the numbers of books he has in his house, and his weight. With each example, Surowiecki showed that the average group answer was more accurate than most individual answers provided, thereby proving (at least to him) his book's central point.[16] The book's core argument works as a parlor trick for the audience, but in reality does the crowd's average answer always lead to the best problem-solving outcome? The crowd wisdom process may work for estimating weights but not necessarily for solving a health crisis or interpersonal problem, or a wicked or ill-structured problem.

Diving deeply into these terms' mechanisms and the extensive research behind them is beyond the scope of this book. While I do not specifically explicate how to better support collective intelligence versus distributed cognition versus collective cognition versus social interaction in general, these terms and distinctions are useful when designing a knowledge game. For instance, questions could be considered, such as how might people work together or apart, competitively or cooperatively, when contributing to knowledge production in a game? How might the wisdom of the crowds support or not support your goals? When do you need people working alongside each other in person versus collaborating through a connected game platform?

Research suggests a possible, though complicated, relationship between knowledge production and teamwork. Knowledge creation is social and relies on networks among people to find and produce it, as well as to share and distribute it, trust it enough to establish it, and continue to replicate, propagate, and build on it. For instance, Wuchty, Jones, and Uzzi found that over a period of 45 years, scientific papers had an average of 1.9 to 3.5 authors each. Yet teams are not always necessary for optimal problem solving. Lakhani et al. found that only 10.6% of the InnoCentive solvers reported that they worked in teams to solve problems. Additionally, only 7.5% of the problem-solving winners worked in teams, but there was no significant difference in success between those who worked in teams and those who did not.[17]

A potential connection between successful knowledge creation and social interaction exists, but what are the conditions and constraints that might suggest more effectively designed and used knowledge games?

Social Interaction Principles

Acknowledging a possible relationship between knowledge production and social interaction leads to further questions. Should people work together to solve problems, and if so, under what conditions? Which game elements might support social interaction in knowledge games? Why and when is cooperation or collaboration, or even competition, beneficial? There are few studies specifically on social interaction in knowledge games, so to begin to answer these questions, I have pulled key ideas from research on games in general, crowdsourcing, citizen science, psychology, and learning. What do we know about social interaction and knowledge making that we can potentially apply to games?

People can learn well in groups and from other people, and games can support communal and social learning. In many ways, learning and knowledge are two sides of the same phenomenon, as learning is necessary for problem solving and knowledge building. There are many theoretical frameworks that support the social aspects of learning, including social constructivism, social learning theory, and situated learning theory, to name just a few.[18]

Bandura's social learning theory suggests that children pick up behavior through observation, even by discerning symbolic representations of that behavior through media and other mediated environments. Based on this theory, it could follow that games can provide an environment where kids and adults learn through the observation of other participants. For example, in *Way*, an online game, two anonymous, but real participants need to work together, simultaneously, to nonverbally communicate what to do next in the game. One task in *Way*—climbing a series of platforms— requires one participant to gesture (using her avatar) where the invisible-to-the-other-player's platforms reside. The other participant then needs to read his partner's cues and respond by jumping on those platforms so he can unlock a door and they both can reach the next goal. The game is collaborative, but no verbal communication (or even text-based chatting) is enabled.

Instead, players need to model behavior to each other through their avatars to teach the other what they need to do next.[19]

How do the theories about learning and social context get translated into games? For one, games often have the ability to support community and social interactions in a variety of ways. In games, participants can interact in a forum, chat, or other community-based platforms. As described further in chapter 7, communities of practice, affinity spaces, and other social configurations can emerge from within or around a game (such as *World of Warcraft*), a game event (such as a *League of Legends* live event), a series of games (such as players discussing the *Walking Dead* series), or even a gaming issue (such as Gamergate) or context (such as games that teach history).[20]

Games are also "emergent cultures" where practices, norms, and values emerge in the community, group, or organization that forms around (and perhaps within) the game. Salen and Zimmerman describe how social interactions around gameplay can emerge from the rules and communication platforms in and around the game. They argue that players who play a game automatically become part of a larger game community, so even if players are not directly playing together, they are nevertheless all playing the same game. For instance, players can engage in a variety of social behaviors and interactions, such as directly sharing knowledge with each other (in person or virtually), transgressing boundaries others have created, chatting with others online, or creating new content that others can use. Rules and norms become shared implicitly—just like any other culture or subculture. New players need to learn these rules and become acculturated into the community, suggesting that a game community itself is like any other human community.[21] This community can emerge from the game organically or be more explicitly supported by communication platforms embedded in or around the game. Such communities are organic and often evolve beyond what the designers originally intended. Even if designers create a well-defined communication platform for a game community, players may adapt and change it so that it evolves and becomes a pastiche of both user-generated and designer-designated interactions and values. What matters, explain Zimmerman and Salen, is how the game system itself and its elements are "made meaningful by its participants."[22]

Gee and Hayes, like Salen and Zimmerman, label the surrounding context of the game as the "metagame." Players develop what Steinkuehler calls "sociocultural norms and . . . shared practices" through these communities, such as learning customs through guilds in MMOGs like *Lineage* or *World of Warcraft.* Many in-game and out-of-game practices acculturate users, such as problem solving together on a quest, engaging in debates and other social events, or participating on fan sites and blogs.[23] For example, I spent many hours poring over *Dragon Age: Inquisition* techniques and tips on Reddit and game forums to learn everything from how to romance certain characters or what ingredients or materials were needed to create new potions or weapons upgrades. Likewise, *EteRNA* participants can check out the community forums to read an RNA puzzle strategy guide, learn techniques for how to solve a specific mission, or figure out how to handle trolls in the chat window.

Squire explains that part of being a problem solver and knowledge acquirer involves becoming a member of a community and, through this, acquiring the necessary skills to actively participate in it. He argues that "we need to look beyond the game itself and toward the broader cultural contexts in which it is situated." He gives as an example the *World of Warcraft* community, which has worked together to map out the entire game through wikis, forums, and resource websites.[24]

Collaboration, cooperation, competition, being part of a community, and other forms of social interaction are motivating and help people feel like they matter. As alluded to in chapter 4, social interaction itself can motivate gameplay, as well as problem solving, skill development, and learning; however, the interrelationships among these are very complex and difficult to pull apart.[25]

On the one hand, people like to feel they belong, and part of that feeling comes from the process of being accepted and integrated into a community. This suggests that enhanced learning is not the only benefit of a community setting but that the social recognition of being part of a community is itself beneficial. McGonigal explains that people want to share experiences with others, build bonds, and feel like they are part of something bigger than themselves.[26] Moreover, people need to feel like they matter, that they are not on the periphery but are central to the mission of the community and of the game or project.

Moreover, knowledge games, through the shared act of playing the game and participating in the game's real-world problems and challenges, may contribute to players feeling like they are part of something larger than themselves and part of a community that is doing meaningful activities. The shared play experience itself may add to a person's feeling that she belongs, as there may be something comforting about knowing others are out there, playing the same game. As I described in the last chapter, a desire to belong may be very motivating.

On the other hand, social interactions in games may not always be beneficial. For example, van der Meij, Albers, and Leemkuil tested 45 students to evaluate whether collaboration enhances learning outcomes or engagement with a game. They used *Lemonade Tycoon*, a commercial off-the-shelf (COTS) strategy game that typically enables one player to run and manage a lemonade stand and work with variables such as type of lemonade, hiring of staff, and weather conditions. The authors set up the study to compare people playing *Lemonade Tycoon* in pairs to people playing the game individually. Their results suggested that there was no difference in engagement in the game between the collaborative pairs and the individuals. The results also implied that collaboration did not affect individual learning outcomes, possibly because of the lack of depth in the dialogues between the collaborative pairs. However, they found that the knowledge that the partners shared while playing was 20% higher than that of what a solo player expressed, suggesting that there were potentials for collective knowledge making.[27]

Other studies suggest that games can motivate social interaction and concomitant benefits. Bououd and Boughzala created a game that focuses on project managers working together to share resources, manage a budget, and complete tasks, and they found that the game was successful in enhancing the ability of managers to practice collaboration skills and better handle their teams. A study by Inkpen, in which players were either at their own computer or sharing a computer to play a game, found that those players who were collaborating and sharing a computer while playing were much more motivated and had higher learning outcomes than those in the other conditions. The researchers theorized that the main reason for this was because partnered players had an opportunity to articulate aloud their moves, ideas, and hypotheses, which helped reinforce their learning and further engaged them in the experience.[28]

A verbalization or sharing of ideas leading to enhanced learning does not always need to happen in real time or with others in the same room. In an analysis of *CityVille*, an online social game accessed through Facebook, Del-Moral and Guzmán-Duque found that the game can support collaboration among users, skill practice and development, and the sharing of strategies. For example, decision-making and problem-solving skills were enhanced for 67% of the game players studied. Part of what seemed to enhance these skills was the use of communication platforms (e.g., forums, Facebook) where players can share information, tips, and techniques. Moreover, the design of *CityVille* encourages players to communicate with each other. The game incentivizes you to help your "neighbors" (fellow players in your social network), such as by constructing a new building, because when you help others you can also benefit. Conversely, a *CityVille* player depends on others to complete these tasks to move ahead in the game, and there may be an implicit expectation that if you help your friend with her tasks, that then she should reciprocate. Because players are so interdependent—they cannot complete the game's tasks without other people—it reinforces the need for social interaction. (Only a superficial level of social interaction is needed to play *CityVille*, however, including posting and clicking on others' Facebook pleas for help, rather than direct, collaborative problem solving). Accordingly, Del-Moral and Guzmán-Duque found that *CityVille* players practiced and developed interpersonal skills (61% of the game players studied reported this).[29]

Even competition with others can be highly motivating; however, this is mediated by a number of factors. First, the possibility of winning cannot be completely out of reach for players. Darvasi explains that when he used his *One Flew over the Cuckoo's Nest* alternate reality game in the classroom, he noticed that those at the top of the leaderboard (or having the highest points in the class) were highly motivated to keep earning more points. Those at the bottom, however, felt that there was no chance of winning, and they became much less motivated to keep playing as a result.[30]

The characteristics of the game player also matter. In a study of exergames, which are games with a physical exercise component, by Song et al., the findings suggested that a competitive context enhanced intrinsic motivational elements for those exergame players who were rated as competitive people but altered mood and lessened motivation for those who were rated as less competitive people. When comparing collaborative to competitive elements in a game, however, they may be both motivating. In a study on an educational math game, Pareto et al. found that there were no differences in motivation between collaborative and competitive activities.[31]

Games can situate social learning. Lave and Wenger suggest that learning and cognition are situated, that is, the context where learning takes place (and is later applied) is central to the learning itself. This situated learning theory suggests that working on real-world problems, with real-world experts and colleagues and within authentic contexts, can be beneficial to learning and

 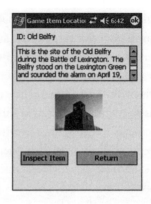

Reliving the Revolution, a mobile, location-based game, was created in the mid-2000s for the Palm Pilot. The goal was to figure out who fired the first shot at the Battle of Lexington. Evidence and documents were triggered to appear on players' mobile devices depending on their physical location in Lexington, Massachusetts. Source: Karen Schrier

applying that knowledge. Wang et al. explain, "In other words, while theoretical knowledge provides a foundation, the insights and skills developed through authentic practice can lead to more meaningful learning."[32]

For example, I created *Reliving the Revolution*, a location-based mobile game where participants need to solve a history mystery and figure out who fired the first shot at the Battle of Lexington. Depending on where participants stand in the actual Battle of Lexington site (the Lexington, Massachusetts, town square and park), images of different virtual historic figures (e.g., Paul Revere), items (a Minuteman soldier's hat), and buildings (Buckman Tavern, the Minuteman soldier headquarters, which is still present at the site) pop up on players' devices, signaling them to query further for a testimony about what happened at the battle. In a study, the game was performed by three different groups of kids and college students. After walking around the Lexington battlefield in teams, playing the game from a specific role, and experiencing the differing testimonials, participants all came together and deliberated who they thought fired the first shot—a conclusion that was different each time the game was performed. The results of the study suggested that the authentic location of the game and the use of real documents and testimonials from people and objects involved in the Battle of Lexington, helped to situate the learning of a historical moment and motivate the practice of historical-thinking skills, engage the players, and make the experience more meaningful, accessible, and applicable.[33]

People can share knowledge and teach each other through games. Sharing knowledge is essential to building new knowledge together. We have seen how players of *EteRNA* or *CityVille* might help each other by sharing tips and tricks in forums. Players can tutor each other through the game, or virtual tutors can participate in the process. Similarly, Steinkuehler and Oh describe how guild members in *World of Warcraft* share everything from special weapons and armor to game strategies and maps.[34]

Peer-to-peer tutoring and the sharing of expertise helps to support overall problem solving and knowledge making in games. People with enhanced or specialized expertise can teach or tutor others, helping them to build the knowledge needed to play the game successfully. This is similar to the cognitive apprenticeship model, part of situated learning theory, which emphasizes how experts can make implicit learning (such as how to fight a big boss battle) explicit by providing assistance to novices and other learners. Experts use several strategies to teach others, such as through coaching, modeling, direct articulation, and reflection on personal blogs or forum posts. In games, some players may even publicize their expertise in the form of badges, ratings, or "likes." This is, in essence, a way of building "social capital" on a site or in a game.[35]

Steinkuehler and Oh argue that in-game learning is even shaped more through apprenticeship relationships than through more didactic or overt instructional methods such as the instructions designed by the game creators. They investigated MMOGs, such as *World of Warcraft* and *Lineage I* and *II*, to understand how apprenticeship works in these games. Their discourse analysis suggests that the "learner" player will identify him or herself as the learner when socially exchanging with another player. Also, players will work together to encourage each other to practice and to provide feedback to each other's gameplay. Experts will model appropriate game behavior for the learner.[36]

To support experts and learners working together and collaboration among people with varied expertise, games can potentially "jigsaw" problems. Jigsawing means that each person takes on a different role or has different responsibilities in a problem context, so they are each more interconnected and interdependent.[37] In *Reliving the Revolution*, for example, I jigsawed the problem of who fired the first shot at the Battle of Lexington by ensuring that each team had information or objects that the other teams did not, meaning that they all needed to rely on each other to share and evaluate that additional information.

Thus, sharing knowledge among people is essential to successful collaboration and integral to knowledge becoming established. Stahl and Hesse describe four paradigms for sharing knowledge:

1. *Sharing mental representations*: Each person has similar mental representations and can explain and compare their representations to achieve sharing but not necessarily change those ideas.
2. *Sharing an object*: Each person has the same object, tool, or goal and can share knowledge of it.
3. *Sharing a situation*: Collaborators are in a common context, environment, or problem space.
4. *Sharing a community*: Shared knowledge comes from being in the same community or society.[38]

Having multiple, diverse views and perspectives can be effective, and games can support their inclusion. Many unsolvable problems are ill structured and complex. Under some conditions, having many diverse opinions and perspectives from people of different backgrounds can be effective for problem solving. Complex and ill-structured knowledge game problems may require different types of expertise and many people to solve them. Hung argues, "Most real world problems are complex . . . [These] problems exceed the cognitive capacity of any individual and therefore require a team of members who possess different but complementary expertise in order to solve the problem."[39] The reasoning behind this notion is that the more interpretations, ideas, and perspectives and greater heterogeneity of experiences or sociocultural contexts you have in a group of players, the greater the potential for successful outcomes and the greater the ability for that team or group to handle the complexity of the problems. Many people playing a game may make better decisions collectively because they are able to fill in each other's gaps, whether in experience, expertise, or perspective on life.

But why would teams or groups in particular be helpful in solving complex, ill-structured problems? One reason that large-scale crowd interactions are beneficial is because of the aforementioned idea that the group's abilities represent more than the sum of its individual members' abilities. Thus, instead of just a collection of individuals working on problems, in the game they become a dynamic system. This relates back to collective intelligence and Surowiecki's wisdom of the crowd concepts, which describe how larger groups of people can collectively make better decisions than individuals can because a group should be able to overcome the knowledge gaps and biases of its members.[40]

With more people, more ideas can emerge, and people can make interconnections among those new ideas. The sum of idea generation becomes greater than its individual parts. This is the notion behind initiatives such as Dell's IdeaStorm, an online platform where people can submit ideas, vote on them, and help them get implemented. However, we need to make sure to avoid groupthink and the "tragedy of the commons." How can smaller voices emerge and become meaningful? Knowledge games may support the emergence of these "longer tail" voices by enabling alternative and even transgressive play,[41] supporting multiple pathways through a game and rewarding participants for contributing novel or unique ideas. Knowledge games can enable a variety of opinions by intentionally inviting and attracting diverse players and encouraging those players to feel comfortable sharing their opinions and providing different views. People may be more apt to open up, share, and participate if they are playing a role in a game, taking on a new identity, or feeling that they are a part of a community that is doing meaningful work, all experiences that knowledge games can support.

Games can encourage argumentation and the consideration of multiple perspectives. The encouragement of many voices, backgrounds, expertise, and perspectives is effective for problem solving—as is supporting the deliberation and argumentation around differing points of view. Klein argues that an argumentation approach may even reduce potential issues with collaboration. This may seem counterintuitive, since we often connect argumentation to ineffective conflict or a debate where only one side wins. In Klein's usage, however, an argument is seen as a process where multiple people are in discussion with each other, constructing points of view and critiquing them. This is distinct from a debate, where people take sides or try to convince others of their perspective. Rather, through argumentation, both sides can explore different viewpoints and integrate them into their own.[42]

In this way, argumentation becomes more of a dialogic process, what E. Michael Nussbaum calls a "critical, elaborative discourse." It is critical, he explains, in that people need to consider, evaluate, and potentially integrate others' points of view and elaborate on misconceptions and misalignments to achieve truly appropriate solutions.[43]

Thus, rather than being an obstacle, argumentation in this sense can help us get to more effective solutions, particularly when we engage in

In *Mission US: For Crown or Colony?*, participants play as Nat Wheeler, a Boston-based printer's apprentice during the American Revolution. Players interact with different virtual characters and complete tasks while learning about the time period. Source: WNET / Channel 13

collaborative argumentation. In science, people engage in argumentation to work through differences, share perspectives, and construct better solutions. Likewise, the Deliberatorium is a technosocial platform that uses research on argumentation theory and social interaction, and which could be applied to everything from political party deliberation to climate-change problem solving.[44]

So does having access to and deliberating among more perspectives lead to better outcomes? Part of the impetus for the mobile game *Reliving the Revolution* was to include multiple diverse perspectives that the players could deliberate over. The act of experiencing and sifting through these different voices helped the participants also understand that any insights gleaned from historic moments are complex and multifaceted rather than one-dimensional or black-and-white. The game helped players conceptualize that history is not a written, established, or linear narrative but a collection of many voices and interpretations that can transform over time. This understanding helped participants to also consider how newer events, such as the Iraq War, might be inscribed differently in history books published in Iraq and in the United States.[45]

In one of the modules of *Mission US: For Crown or Colony?*, game players experience the Boston Massacre from a variety of perspectives. Each participant receives a different set of four "vignettes" of what happened, which helps players understand that there are multiple perspectives on and interpretations of historic moments such as the Boston Massacre. Source: WNET / Channel 13

The supposition that deliberation supports learning and problem solving also helped to inspire *Mission US: For Crown or Colony?*, a game created for middle-school social studies students, and particularly a module in the game on the so-called Boston Massacre. In the game, participants experience the colonial era in Boston through the eyes of Nat Wheeler, their avatar, who is a printer's apprentice in colonial Boston. At one point, the players stumble on the chaotic Boston Massacre scene and view vignettes about what happened. Each participant receives a randomized set of four vignettes, or possible illustrations of the incident, so that no two sets are the same. In the classroom, where the game is typically played, participants can then deliberate (aloud or through journals) what happened from their own perspectives, share their opinions with other students and learn their perspectives, and consider why there were so many different ways to view the Boston Massacre. In the next leg of the game, game players are asked to make decisions about what happened in the Boston Massacre, and their choices affect the outcome of their game. Research conducted on the game suggests that the participants practiced historical empathy and other critical skills, such as evaluation of evidence and formulation of arguments, all skills integral to problem solving and decision making.[46]

Research also supports the use of argumentation to make more effective decisions and solutions. Nussbaum describes something called the "two wrongs make a right" phenomenon. This occurs when two people who both have distinct, inaccurate solutions to a problem talk through their differences. Often, Nussbaum argues, this type of communication leads to a more accurate, communally developed strategy or solution. In part, this may be because when people are challenged by others, it encourages them to be flexible and not fixated on one way of solving a problem. This is not to say that the sharing of different perspectives is still effective if it becomes contentious. Keefer, Zeitz, and Resnick found that students who were engaged in small group discussions of literature and had noncontentious, collaborative argumentation had more enhanced understandings of the literature's themes than those groups with adversarial argumentation.[47]

Argumentation supports learning and skill practice, but does it also encourage knowledge production? Some research has suggested correlations between deliberation and argumentation in games and more effective knowledge production. One way this may occur is through the sharing and deliberating of perspectives, opinions, hypotheses, and so-called facts in games, which can help enhance a community's overall accuracy. Outcomes and solutions are higher quality because people are double-checking and fact-checking each other or providing alternate perspectives or possible answers. Games can encourage fact-checking by enabling multiple participants to do the same tasks and then compare the results. *Reverse the Odds* asks many players to look at the same cells and judge them, and then it compares the responses; Galaxy Zoo invites multiple people to categorize the same galaxy images. Thus, knowledge games can inspire participants to view and replay each other's activities and provide feedback on it, either through external mechanisms or within the game world itself, such as through multiple playthroughs by different people.

Also, games can enable a group's testing and retesting of hypotheses. For example, Choontanom and Narid describe the process of theorycrafting in games such as *Starcraft* and *World of Warcraft*, where people work together to generate ideas about what is happening in the game, test each other's hypotheses, mount arguments, collect and analyze relevant data, post interpretations and results, and engage in debates about the findings and what they mean—similar to how, for example, scientists and other researchers work together. The authors argue that while textbooks and other educational materials often do not incite active participation, games can enable players to put forth possible theories and revise or question each other's ideas and assumptions. In their analysis, theorycrafting, which is supported through the community around an MMOG, encourages community vision and revision, contribution and questioning, and deliberation and argumentation, which all potentially lead to more effective knowledge production.[48]

Thus, research suggests that multiple, distinct expertise and perspectives should be shared and deliberated to provide meaningful and effective interconnections, novel solutions, and knowledge building.[49]

A game can also support reflection on its emergent preconceptions, as well as consideration of players' own identities, inner biases, and limitations. Squire explains that part of participating in a community is not just being in it but constantly evaluating it and reflecting on it to understand its biases, limits, and potentials.[50] He uses as an example his research on *Civilization*, where students were able to see how his Civ editor modeled different variables, such as population growth and happiness, and explained the logic behind the rules.

Why is it important that players lift the hood and reflect on the community and the game itself? Turkle suggests that it is useful to expose the rules and algorithms behind a simulation to better understand that simulation, even just so that we understand that it is purposefully *designed* and in part *emerges* from the actions of players and their community.[51] It is important for players to understand that the game could have been designed many other ways, and the way it is designed affects the questions we ask, the knowledge and other outcomes that are produced, how we approach and solve problems, and how the knowledge is then shared with and adopted by others. Moreover, how the play and culture within and around the game organically and dynamically emerges is often unpredictable and may incorporate its own set of biases, predispositions, and limitations in how knowledge is accessed, produced, shared, and established.

Beyond reflecting on the system of the game and its emergent properties, we also need to consider our own identity within the game and how each individual player affects, and is affected by, this dynamic system and helps it evolve. Gee considers how games are semiotic systems and designed creations meant to engage people in, for example, exploring new identities (such as via an avatar or main character), new environments, and new challenges. He discusses the idea of real, virtual, and projected identities, which, represent, respectively, one's identity in real life, one's identity as the virtual character or avatar in a game (or even as oneself in the game world), and a projected identity, or a hybrid identity in which your real identity is projected to some extent onto that of the virtual character.[52]

In a knowledge game, participants can potentially project their own real-world identities onto a game identity, becoming a game player, community member, problem solver, and scientist, social scientist, or humanist in their own right. Players are not just role-playing as problem solvers, they *are* the problem solvers, and they inscribe that role with their own history, values, experiences, and sociocultural contexts, which include both their own identities and the identities taken on in the game and real world.[53]

Knowledge games may even enable you to experiment with and explore your own identity or even someone else's, potentially freeing you from your own preconceived notions, misconceptions, and perspectives, and enabling you to try on new roles. When people play new roles—regardless of whether they are trying to "act exactly like themselves" or pretend to be someone else—they are still stepping outside of their everyday world and into a new exploratory realm. This act of taking on a role may help someone reflect more on his own values, circumstances, and real-world issues, and it may help him avoid being stuck in his own perspective or sociocultural lens, which also may lead to more effective knowledge making, as well as greater transparency and reflection on the ways in which this knowledge is socially constructed and what it really means.[54] Moreover, unlike other platforms for problem solving, knowledge games can potentially attract many different perspectives on a problem, leading to more possible solutions, or a solution with a broader reach, which takes into consideration the needs and desires of many more people.

When Is Social Interaction Less Appropriate?

Social interactions such as collaboration are not always essential, preferable, or valuable for effective knowledge production. Thus, when designing or using a knowledge game, we first need to decide and reflect on whether purposefully designing for social interactions, such as collaboration, competition, or cooperation, or the generation of communities within or around a game, is appropriate for the needs and goals of the project, as well as the audience, type of problem, and knowledge we want to produce.

As of yet, few studies focus on why perfectly well-meaning social interactions and collaborations fail or are nonproductive. Most studies instead have considered more fully the adverse aspects of social interactions, including "exploitation, negative impacts on students, and unethical behavior," all of which can problematize knowledge games as well.[55] For instance, players can be exploited for their hard work in a game (a topic that is discussed further in chapter 7) or they can purposefully band together to rig a game with erroneous data to subvert knowledge production and sharing, similar to the way police officers were marking incorrect police traps on the Waze app (see chapter 1).

But why does communication sometimes break down? Why are interconnections among people not always effective? Certainly we know of many college students working in project groups who complain about those groups' dysfunction, and the sum is not always greater than its parts. When I was studying the collaborative game *Way*, some of the player pairs were successful in completing the game together, but in other cases one of the participants dropped out (meaning that both players automatically failed the game and had to restart it). Klein points to problems that are present in communication platforms. He first evaluates time-centric organized communities, such as e-mail or Web forums, where people can see the most recent threads or e-mails first, for example. Although knowledge games have not been studied directly in relation to time-centric communication, Klein's findings might be applicable. These include:

1. *Scattered content*: It is hard to focus conversations and issues.
2. *A low signal-to-noise ratio*: It is difficult to distinguish the novel and unique conversations to attend to over the din.
3. *Balkanization*: There is little time spent going to oppositional threads, so you do not see a cross-pollination of ideas.
4. *Dysfunctional argument*: There are more bias-based than evidence-based arguments.

Klein explains that there are also problems in topic-centric platforms, including *redundancy*, or lots of ideas that are similar to each other, and *non-collaborativeness*, or many ideas put forth but few that people are working on together with complexity.[56]

Thus, even if social interactions are beneficial to a knowledge game, there are many barriers to enhancing interactions among people. One barrier to social interactions is that access to a game and equity to access is never guaranteed for all potential players (this is discussed further in chapter 7). Games need to attract a diverse audience at all points in the process, from recruitment and motivation to play, to sustaining play and embracing people to engage with the game's community. Even implicit or subtle messages can affect how a person decides whether to participate in a game or to continue to participate. Who can participate and how they can participate will also affect how players socially interact and the knowledge a game community produces. Of course, these issues are not just related to games but to all facets of human life.

In chapter 7, I discuss the landscape of participation in more depth. In the next chapter, however, I address the history of amateur participation, how it might also affect participation in knowledge games today, and the acceptance of knowledge produced through it.

PART THREE

PERSPECTIVES, POTENTIALS, AND PITFALLS

CHAPTER SIX

Amateurs

In fifth grade my friend got an exciting new piece of technology for her birthday. It was a telescope.

What's more, it was a telescope that took photos! We spent that summer and fall watching the stars, recording the night sky, and tracking the movements of stars in constellations such as Cassiopeia. The telescope's camera was able to capture how the stars moved using time-lapse photography, which meant that we got some great photos with streaks of light across it. Up high and in the images we created, we watched each night as the "W" of Cassiopeia's stars danced through the sky—stars we adored just like any of the other celebrity "stars" of the day. We were fan girls of Queen Cassiopeia, eventually entering our data in the school science fair.

Little did we know that we had entered a world of amateur stargazing and world-gazing that has been happening in different forms for thousands of years.

Similarly, one night in 2010, Josch Hambsch stepped out to his backyard, as he had done most nights for many years, and observed the star GV Andromeda with his CCD camera. Hambsch, a Belgian member of the American Association of Variable Star Observers (AAVSO), first got involved in monitoring the sky after he observed a gamma-ray burst in 2003. Since then, he has logged more than 150,000 observations. Hambsch's monitoring of GV Andromeda's brightness was then submitted to the AAVSO's database in 2011, making it the 20 millionth observation of a variable star done by members of the AAVSO since its founding in 1911. These completely volunteer night sky observations, conducted by over 7,500 amateur astronomers, have amounted to a whopping 1.67 million hours of observations and $27.5 million worth of donated time. By late 2013, the observations in the AAVSO database had surpassed 24 million.[1]

The AAVSO has been a highly organized, social, prolific, and active-amateur-led knowledge-producing ecosystem for over 100 years. Amateur astronomers today can also get involved with Galaxy Zoo, an online collaborative platform where more than 200,000 volunteers have classified

The American Association of Variable Star Observers (AAVSO) is an organization that supports, collects, and analyzes amateur contributions of star observations. It has been active for over 100 years. Source: Courtesy of the American Association of Variable Star Observers (AAVSO)

galaxies in photographs "taken automatically by a robotic telescope [as part of the Sloan Digital Sky Survey], and have never before been seen by any human eye. You can think of Galaxy Zoo as a cosmological census, the largest ever undertaken, a census that has so far produced more than 150 million galaxy classifications." What would have taken years to accomplish can be done, instead, in a few weeks. The Galaxy Zoo website explains that "within 24 hours of launch we were stunned to be receiving almost 70,000 classifications an hour. In the end, more than 50 million classifications were received by the project during its first year, contributed by more than 150,000 people." Subsequent versions of this platform include Galaxy Zoo 2, launched in 2009, which has obtained over 60 million classifications; Galaxy Zoo Hubble (2010), which invites users to help compare galaxies of the past to galaxies of today; and Radio Galaxy Zoo (2013), which involves annotating images of potentially erupting black holes that could be dragging jets of material into them.[2]

With Galaxy Zoo, amateurs have discovered an entirely new class of galaxies, called "Green Pea Galaxies," and observed the first "quasar mirror, an

In Galaxy Zoo, a citizen science project, participants assist in classifying galaxies, such as by noting its shape, using images taken by high-powered cameras and telescopes. Source: Zooniverse

enormous cloud of gas tens of thousands of light-years in diameter, which is glowing brightly as the gas is heated by light from a nearby quasar." Between its inception in 2007 and March 2015, the Galaxy Zoo classifications contributed to 56 scientific papers.[3]

But it is not only star-struck amateur astronomers who have made, and continue to make, contributions to public knowledge. Today there are many thousands of citizen science projects, "ranging from breeding bird atlases to aquatic insect counts, from frog-watching projects to reef fish surveys," according to Dickinson and Bonney, researchers at Cornell's Ornithology Lab, which hosts the eBird and FeederWatch programs.[4] Whereas distributed computing projects such as SETI@Home use spare processing and computer power, efforts such as Galaxy Zoo and the activities of the AAVSO use spare human power.

Citizen science and related activities did not arise out of nowhere. In this chapter, I take a big step back and explore the history of amateur participation in science. Perhaps we can look at the history of citizen science to help us conceptualize the future of knowledge games and their role in problem solving and knowledge production.

Citizen Scientists Unite

Citizen science is a relatively new term for a very old phenomenon, one that even Aristotle alluded to: "any person can participate in scientific research—regardless of background, formal training, or political persuasion." Citizen engagement in science, even in large, organized, supportive, and distributed communities, has had a long history.[5]

One of the longest-running citizen science events is the Christmas Bird Count, which was initiated at the turn of the twentieth century by the National Audubon Society. Each year, since 1900, tens of thousands of birders brave the cold, counting birds in late December and early January and contributing their findings to the Audubon Society's database. This bird information helps to model North American bird populations over the long term and can even help flag environmental problems or threats. According to the Audubon website, data from the Christmas Bird Count helped to record a decrease in the American Black Duck in the 1980s, leading to tighter hunting regulations. Information from the count also facilitates our estimation of climate-change factors and effects. The aforementioned FeederWatch, an organized bird-watching program started in 1986 and run by Cornell's Ornithology Lab and Bird Studies Canada, literally feeds into programs like the Christmas Bird Count. Volunteers pay a small fee and are sent a research kit to participate in the project. They set up their own feeder and use their own birdseed. Then, from November through April, they count the birds and identify the bird types that visit their feeders, sending the data back to the FeederWatch organizers. The program's over 20,000 volunteers have been instrumental in tracking bird species numbers, ranges, and distributions throughout the North American winter for the past 30 years.[6]

Then there is botanizing—the practice of collecting, discussing, preserving, identifying, socializing, and reading about plants, conducted by amateurs without any formal training. In the 1800s, and especially between 1830 and 1890, tens of thousands of botanizers were involved in the science community, forming "the majority of most botanical societies, and in many learned and scientific societies, . . . [and the] legs, hands, and eyes of individuals and institutions," according to historian Elizabeth Keeney.[7]

In fact, many of the people we think of today as the genius scientists or naturalists of the past—Thomas Edison, Michael Faraday, Henrietta Swan Leavitt, Charles Darwin, Henry David Thoreau, Thomas Jefferson—were

In Project FeederWatch, participants set up bird feeders and count how many birds of different types visit it. This map shows the locations of North American participants in the 2014–15 season. Source: http://feederwatch.org/PFWMaps/participants

actually considered amateur scientists in their time. Famed amateur scientist Forrest Mims explains, "Contemporary science has its roots in the achievements of amateur scientists of centuries past. Although they lacked what we would define as formal scientific training, they deciphered the basic laws of physics and principles of chemistry. They invented instruments. And they discovered, documented, sketched, and painted planets, comets, fossils, and species."[8]

The Rise of the Scientist

How has the participation of amateurs in science changed over time? Dickinson and Bonney argue that we need to rethink the currently shifting relationship between citizen and scientist. Computing pioneer Michael Nielsen contends that we are at the cusp of paradigmatic changes in the practice of science and scientific discovery, due in part to networked technologies and the new collective cognitive problem-solving tools and activities those technologies encourage.[9] Is the practice of science, and knowledge creation in general, in transition? Are we experiencing a shift in the perception and practice of science due, in part, to broader cultural, economic, political, and social changes, as well as to citizen participation in citizen science, citizen games, and other emerging participatory platforms and tools? What is the role of knowledge games in this shift?

This idea—that any citizen can (and should) participate in science or in other "professional" pursuits—has been complicated. "Amateur" is a dirty word today, but it was not always that way. Amateur's Latin root progenitor, *amator*, means "lover, or lover of, or devotee." The word was a neologism in the eighteenth century and initially referred to people who wholly and enthusiastically love what they do. Today, however, amateur connotes a dabbler or dilettante; someone who does something as a side hobby, has less knowledge or know-how than others, or even works against the profession. When we hear that someone is an amateur scientist, we do not typically think of that person as making a valuable or trustworthy contribution. Rather, we think of amateurs' activities as only personally useful, and even inferior to those of professionals.[10]

How did the schism between amateur-scientist of yesterday and scientist-professional of today happen? What implications does this shift have for amateur participation in knowledge games and other emerging platforms or practices? To more fully understand this transition, we need to take a journey back to the broader cultural, economic, and political changes that happened centuries ago, to see how they affected the relationship between amateur and scientist. We need to understand how science became a profession in its own right.

In the seventeenth century, science was seen as a calling, and it thereby had an almost sacred quality and moral responsibility. By the turn of the century, however, the perception of the scientist was in limbo, and Steven Shapin explains that "the identity of the scientist was radically unstable. To be a scientist was still something of a calling but it was becoming something of a job; it was still associated with the idea of social disengagement but increasingly recognized as a source of civically valued power and wealth; it was still associated with a notion of special personal virtue but it was on the cusp of moral ordinariness."[11]

So what happened? Shapin and Nielsen point to a complex web of economic and social changes that altered the perception of science as a public good and as "materially necessary" for society. Science became increasingly valued throughout the eighteenth and nineteenth century, and the tangible social and economic value of scientific work helped to integrate science institutionally and ratify it as a distinct occupation and division of labor.[12]

The increased value given to the practice of science was due, in part, to its more obvious public benefit, particularly following the scientific discoveries of the seventeenth century, and to "prestige conferred on leaders (such as the Medici) by association with such discoveries."[13] These leaders would then, in turn, financially subsidize future scientific activities. The perceived necessity of science continued to grow with the shift from an agrarian to industrial workforce, as there was a need for specialized, technically savvy workers and scientific instrument builders.[14] Consider, for example, the clothing industry. Prior to the Industrial Revolution, clothes were typically hand sewn and knitted by family members or local community members. With the advent of new technologies to help create and distribute clothes, such as the sewing machine, people were able to produce more clothes and sell them to a variety of consumers farther away. The mass production of clothing enhanced the need for clothing dyes, which could be used to distinguish and differentiate clothing so that people did not feel they were all wearing the same thing. The manufacturers turned to chemists, who helped synthesize new colors or longer-lasting dyes. The demand for chemists in turn contributed to the growth of science specializations in academic institutions, such as in the field of chemistry in Germany.[15]

Cultural and social changes in the profession of and practice of science had been percolating even prior to the Industrial Revolution. Historian Elizabeth Eisenstein points to the influence of "print culture" following the invention of moveable type by Johannes Gutenberg, as helping to shape scientific activity. Scientists and amateurs, now wielding printed books, were able to compare what was written in the books to their own observations of the world and even to supply their own corrections. This is similar to how current scientific journals and periodicals work today, with researchers reviewing and vetting each other's work. These practices, coupled with the ability to print and distribute large volumes of texts, contributed to, in part, the advent and requirement of scientists and other researchers to publish findings through a rigorous vetting process.[16]

A desire to publicly disclose results to the community rather than remain secretive in one's scientific endeavors, as well as a greater emphasis on accuracy and data preservation, also contributed to changes in the evaluation and institutionalization of scientific activities and the way they were culturally assessed. Nielsen explains that prior to the mid-seventeenth century, scientists were "reluctant to publicize their discoveries in any way at all." There was no incentive for them to share findings—in fact, sometimes there were deterrents. Nielsen tells the story of Galileo, who in the early 1600s "made the mistake of showing a military compass he had invented to a young man named Baldassare Capra . . . It took Galileo years of effort and considerable expense to regain the credit for his discovery, not to mention his reputation." When, in 1655, the first scientific journals started, they emerged from and signaled an increasingly open, sharing-focused culture among scientists. Both the public's need for sharing scientific knowledge and the patron's desire to gain public approval for funding any scientific research "were best served if discoveries were widely shared through a medium such as a scientific journal. As a result, patrons demanded a shift toward a scientific culture in which it is the sharing of discoveries that is rewarded with jobs and prestige for the discoverer."[17] Thus, the dynamic social, interpersonal, and psychological interrelationships among scientists and the public shaped and were shaped by the scientific knowledge that was produced and how it was accepted, shared, and established.

The Fall of the Amateur

Where do amateurs fit into the changes in the scientific profession? Initially following the advent of print culture in the mid-1500s, amateurs were still very much part of the scientific conversation and able to influence the practice of science and creation of knowledge. Eisenstein writes that "once fresh observations drawn from nature could be duplicated in printed books, they became available to scattered readers who could, at long last, check books against nature and feed back corrections with new observations to be incorporated into later editions."[18] Those readers, many of whom were amateurs, were called on to help check and double-check against the natural world. In other words, these reader-amateur-scientists were part of a larger community where the practice of scientific tasks (such as collecting plant specimens or double-checking findings) were shared among them, and any information was then communicated back to each other through revised editions of printed scientific texts.

For example, many amateur botanists lobbied to contribute to Mattioli's series of books (commentaries on Dioscorides's original systems of plants in *De materia medica*) in the 1500s, trying to get their individual interpretations and discoveries of new plants added to revised editions—and thereby to be "immortalized." Thus, through the community surrounding his publications, Mattioli accessed the "collective intelligence" and "distributed cognition" of amateurs, gathered their interpretations, and deemed them as valuable. Later editions of Mattioli's books were revised and corrected to "take into account this feedback from readers, [acting] very much like later scientific periodicals."[19] Functionally, this also seems quite similar to how crowdsourcing, citizen science, and knowledge game communities today work together to contribute to knowledge of birds in the FeederWatch project, stars in Galaxy Zoo, or cancer cells in *Reverse the Odds*.

But by the late 1800s, amateur scientists such as the botanizers were no longer considered a vital part of the scientific community and were being relegated to its peripheries or completely excluded. Prior to the late 1800s, amateurs had made up a large proportion of many of the scientific societies.[20] By the turn of the century, they were completely left out.

Why were amateurs excluded from the scientific community? Part of the reason is because as science became increasingly valued and institutionalized, its inner social mechanics and mechanisms also transformed. By 1900, amateurs were no longer part of the collaborative conversations or social exchanges among scientists, even though both professionals and amateurs together had been participating through the communication platforms (such as books or in scientific societies or at social gatherings) that facilitated these exchanges, shaped scientific practice, and made lasting contributions to knowledge. At the turn of the twentieth century, people who wanted to take part in scientific knowledge-making conversations needed to demonstrate their mettle by their training, peer review and approval, or work experience. In other words, they needed to prove that they were not amateurs anymore. The people who did not meet these criteria were essentially kicked out of the scientific social circles and social definitions of "scientist." Likewise, because amateurs were less and less able to contribute to official knowledge production, and therefore, not contributing as much, they were increasingly seen as dabblers, whose contributions were irrelevant. Eventually, the schism between the amateur and professional scientist was complete.

Which brings us back to today. That certain practices are now called "citizen science" rather than just "science" alludes to the continued distinction between amateur and professional scientist. The title of scientist, according to Eisenstein, refers to "not only his command of technical literature—but also his capacity to put his findings in a form where they can be correlated with prior work—where they can be accepted or rejected by consensual validation—that helps to distinguish the scientist from the shrewd observer or from the speculative 'crank' and the ingenious gadgeteer."[21] The scientist needs to continually prove her worth to be included in the broader scientific discourse by making contributions that others care about and trust—and frequently this is through publishing in peer-reviewed journals or persisting in tenured or institutionally vetted positions. Scientists need to be socially approved. A person cannot just walk into a laboratory and start a restriction enzyme digestion procedure. You need degrees from fancy schools, years of formal training, and mountains of research publications to prove your pluck.

Of Play and Time

The more amateurs were pushed out of the communication platforms that enabled scientific knowledge production, the more they also withdrew, reinforcing the schism. I discussed why professional scientists wanted to separate themselves from amateurs, but why did amateurs simultaneously want to withdraw from the conversations? This requires understanding another important factor—our changing conception and practice of leisure.

Let's first go back to the seventeenth century, when the colonists first arrived in America—tired, hungry, and determined. They had left Europe in search of religious freedom and exhibited their "Protestant work ethic" in full effect, needing to labor hard to survive in the New World: "there was little time, money or energy to support amusements or public entertainment." But by the mid-1700s, as the colonists settled into their new homes and were not as focused on day-to-day survival, their strict policies and religiosity against play and leisure waned, though it still influenced culture.[22]

The tension between conceptions of leisure and hard work continued through the 1800s, when there was an expansion of amusements, such as theaters, drinking, dancing, and sport, but also a revival in the idea of the "honest toil," which tied hard work to morality. In the late 1800s, there was a steep increase in pastime participation, with a cultural climate more open to play and fun. During this era, the amount of leisure time also increased, as the average workweek declined from 69.7 hours per week in 1860 to 61.7 hours in 1890 and then down to 54.9 hours in 1910, due in part to shifts from agrarian to factory jobs following the Industrial Revolution and in part from labor reform policies and the formation of unions.[23]

How economic classes viewed leisure also shifted. The Protestant work ethic and religiosity, coupled with Victorian values from Europe, influenced beliefs that leisure time should be spent in edifying ways "that had moral and social utility," particularly for upper-class citizens. This notion was at odds with urban working-class citizens and newer immigrants, who often felt that fun and frivolity were the guiding attributes for one's leisure time.

Values and morals were ascribed to leisure activities, and these also had class implications. In other words, taste hierarchies were applied to how one spent free time. Mass culture and mass amusements, such as film, emerged not only as entertainment but also as a "threat to traditional morality and values." Moreover, as the distinction between luxury goods and common commodities became increasingly shallower, the upper classes needed other markers to distinguish themselves from lower-class citizens, and part of this distinction was in how people spent their leisure time and described it to other people.[24]

So where does this put amateurs, who mainly pursued science during their leisure time? As the conception of leisure transformed and amateurs were pushed out as central contributors to the work of science, they simultaneously removed themselves from participating in tedious types of activities. For instance, over the nineteenth century, amateur botanizers were becoming less and less interested in the dull and drudgery aspects of collection and analysis and more interested in its fun, escapist, and personally fulfilling aspects, according to Elizabeth Keeney. "Amateurs were also growing ever less concerned about the work-like qualities of botanizing, and more interested in pleasure, relaxation, and other recreational activities." Part of this shift was shaping, and being shaped by the school system and the botanical journals of the day. School textbooks and tasks that were deemed too dull started being skipped in school in lieu of more fun learning activities. Journals targeting amateurs related that "amateurs should botanize only if they enjoyed it. They should not aim toward contributing to science. They should not subject themselves to boring laboratory work. They should focus on field botany and observation . . . Botany was to be a hobby, something one dabbled at, because it was fun" and therefore not serious.[25]

On the flip side, science professionals sought to label themselves as serious workers and not dabblers, and became increasingly motivated to contribute to a body of knowledge that the greater society saw as quite worthy and necessary. In sum, by the end of the nineteenth century, more and more people saw leisure time as time that should be spent on something fun—to some extent still edifying, morally appropriate, and personally enhancing—but decidedly unwork-like. Work and fun became more clearly demarcated, at least in the public's perceptions, just like the distinction between who was a serious scientist and who was a rollicking amateur. Chapter 7 will cover more about the tension between work and fun.

Of Games and Frames

But are the actual activities of the amateur and scientist really so different? Max Weber explains, "Scientifically, a dilettante's idea may have the very same or even a greater bearing for science than that of a specialist. Many of our very best hypotheses and insights are due precisely to dilettantes. The dilettante differs from the expert, as Helmholtz has said of Robert Mayer, only in that he lacks a firm and reliable work procedure. Consequently he is usually not in the position to control, to estimate, or to exploit the idea in its bearings."[26]

Materially and functionally, the amateur and scientist may be behaving similarly, but cultural, economic, and social perceptions matter, as well as how the activities are framed.[27] What happens, then, if amateurs and scientists are working together in games? Do knowledge games subvert our understanding of this distinction between work and play, scientist and amateur—or do they support it? If the actual activities that the knowledge game player is doing are the same, substantively, of that of a scientist data collector and interpreter, what are the implications for distinguishing, or not distinguishing, such activities in a game?

The framing of work versus play, as well as the cultural acceptance of this distinction, is important in understanding when a scientist is a scientist and when she is an amateur and what values we ascribe to the differences.

Gary Alan Fine, building on Goffman's frame analysis, applies frames to games. "Like many social worlds (acting, storytelling), fantasy games produce a 'make-believe' world set apart from the everyday world. By playing fantasy games, participants implicitly agree to 'bracket' the world outside of the game . . . Every play world has a set of transformation rules that indicates what is to be treated as real and how it is to be treated as real within the make-believe framework." These agreed-upon "rules" that Fine refers to help to maintain the separation between scientists and amateurs, even when they are doing the same functions or tasks in a knowledge game.[28] The players tacitly agree to bound the game as separate from real life; as a result, the normative relationships between scientist and amateur remain unharmed.

Likewise, games are typically framed, morally judged, and considered culturally, on the whole, as frivolous, fun, and carefree excursions.[29] This framing of games stems in part from the association of games with children, play, and nonwork activities, as well as the aforementioned Protestant "work ethic" and Victorian conceptions of leisure, among other reasons. Games are viewed as nonserious pursuits; therefore, they do not threaten the serious character of true science. "Fun" is not just a cultural frame; from a practical perspective, it's often a defining requirement and design constraint of games. Designers constantly hear that their games must be fun, first and foremost. The problem with those educational games? They just aren't fun enough. The problem with your failed playtest? The mechanics weren't fun. The problem with reality? It's not fun, so we should make it more like a game—because that's where the real fun begins.[30]

But there is a whole lot of work—and sometimes really hard work—in games.[31] We do not, however, frame games as "work," even though we may actually work more in games than outside of games, or we may learn more, express more, or create more through games. In fact, Ian Bogost posits that the fun in games is the *work*, the hard stuff, and the boundary pushing, rather than "the spoonful of sugar" that's needed to take medicine or the chocolate in the "chocolate-covered broccoli" metaphor. He asks, "What if we arrive at fun . . . by embracing the wretchedness of the circumstances themselves? . . . What if, in a literal way, fun comes from impoverishment, from wretchedness? What if it is in the broccoli without the chocolate?" Bogost maintains that fun is "excreted" from the structure in games—the given shared constraints that we accept as players, that we take seriously, and that we push to their "logical extremes." Thereby, "play turns out not to be an act of diversion, but the work of working a system . . . Fun is . . . a nickname for the feeling of operating that system, particularly of operating it in a way we haven't done before, that lets us discover something within it that was always there, but that we didn't notice. Or that we overlooked, or that we found before and are finding again."[32] Perhaps the fun of games *is* the problem solving and *is* the discovery that occurs by caressing the limits of our systems and institutions. Games, perhaps, are about reinventing, about new perspectives, new solutions, and finding novel possibilities in the quotidian structures of our world. Games, possibly, are *about* knowledge production and about reimagining the limits of knowledge.

Fine contends, "'Fun' would seem to be a *sine qua non* for gaming, but 'fun' is a flickering experience, and along with it flickers engrossment and the stability of the frame." Fun is not maintained steadily throughout a game; rather, it wavers, depletes, and reignites as the person plays the game, pushes its boundaries, discovers new connections, and posits new claims about its system. Perhaps the shifts in acceleration toward these boundaries, rather than the escape from them, are what let us know that we are having fun.[33]

Games, thus, are a bounded, framed, and protected place to reevaluate institutions and institutionalized problems—to push the borders of not only the game itself but also the systems and rigid structures of our world, which can be abstracted and reconfigured, safely, in a game. Sniderman suggests that "people want to believe that their beloved institutions are sacred, unchanging and right, but (almost) no one feels that way about games . . . they can provide an analog to other, more 'important' and more complicated, aspects of life and thus can help us see what otherwise might be invisible" about those institutions, or about any of the inherent perspectives, biases, or rules we take for granted.[34] The fun in games is the ability to seriously push boundaries, which, ironically, you may only be able to do if the game is framed as something not serious and not consequential. This might make games particularly relevant to knowledge production.

But once games are no longer framed as an escape, as a playspace, as make-believe, as frivolous leisure, and as separate from work, what happens? Perhaps, if games are framed as serious endeavors, we cannot safely and without consequence traverse boundaries, subvert, reconfigure, and reevaluate the norms, institutions, and embedded social and scientific systems that we take for granted. If, in the case of knowledge games, they are taken seriously, will no one trust that they are games—and will no one want to play them?

Perhaps it is *through* games that amateurs are allowed to move, at least temporarily, into the role of professional scientists. Because games are framed as fun and separate from reality, as play or make-believe, even if amateurs are, in actuality, doing the same work as a scientist and contributing to scientific knowledge, it's okay because they are game players and therefore still amateurs, first and foremost. The status quo has been maintained.

Knowledge games may be unique in that they simultaneously support the motivations of an amateur—fun and personal satisfaction, social interactions and community, and a desire to contribute to knowledge—while also supporting the needs of professional scientists, such as furthering their social status as scientists, educating the public about science, and producing scientific knowledge. At the same time, the sense that leisure pursuits should be both an escape from reality and a productive use of time still persists, as does the tension between the Protestant work ethic and the conflicting desire for carefree and liberating recreation. Games, and especially knowledge games, might fulfill the need for edifying leisure and resolve this dialectic.

But what if it's more than that? What if knowledge games, and other types of games, elicit what Bogost describes as the kind of "foolishness" required to see what is currently unseen. This foolishness is a serious commitment, a curiosity, and a passion for asking what else is possible, a quality that has been ascribed to game players, amateurs, and all knowledge seekers. What if accessing this foolishness is what is needed for novel problem solving, for discovery, for realigning our institutions, and for solving the problems we need to solve? What if we need game players and amateurs—the fools—to be able to "overcome unnecessary obstacles" that exist in our world?[35]

In other words, games might enable amateurs to traverse the boundaries of propriety and trespass into the sacred world of the professional. They also might allow professionals to let loose, step outside their world, and to poke at established paradigms. Perhaps it is through these mutinous movements that we can more readily excavate new knowledge, innovate, and make real-world change. What implications, then, does this—the sense that knowledge games have a unique ability to marry seriousness and foolishness—have on knowledge creation? Do amateurs and professionals need to be immersed in a playspace to interact, or can they converge on an equal playing ground to do the work of "knowledge making" side by side, within current institutions? Do they all need to believe that their activities are only "half-serious"[36]—in other words, do they need to suspend beliefs, statuses, and identities to be able to interact? In other words, do they need to play a game for the subversion of the professional-amateur relationship to occur?

To further explore this complexity, we need to consider another factor in the relationship between amateur, game player, professional, and society: trust.

A State of Lulz and Trust

What implications does trust have for knowledge creation?[37] Many years ago, people lived in local communities where they knew whose word to trust. The harvesting tips from the local farmer with the juiciest tomatoes were carefully considered, whereas advice from the known town liar was carelessly tossed aside. The perception was that it was easier to vet the truth if you knew everyone face-to-face, especially if you also knew their cow-herder parents, their botanizer brother, and their churchgoing wife.[38]

But something happened as knowledge became more separated from the person with whom it originated, and as we went from oral to scribal to a print culture. We could no longer always look a person in the eye or judge the personal characteristics, materiality, or "urgency" of her script to evaluate its validity.[39] With print, we needed to find new ways to decide what knowledge was spurious and what was useful. The historian Adrian Johns explains, "A central element in the reading of a printed work was likely to be a critical appraisal of its identity and its credit. Readers were not without resources for such an assessment. When they approached a given book, with them came knowledge about the purposes, status, and reliability of printed materials in general—knowledge they used to determine the appropriate kind and degree of faith to vest in this unfamiliar object." When it comes to knowledge creation and public acceptance of this knowledge, trust is essential.[40]

But what about printed scientific knowledge? No longer did a person have to be a direct witness to an experiment, or even in correspondence with the person doing the research; instead, a reader could be a "virtual witness" to a discovery or proof.[41] As discussed earlier, the scientist and his research needed to be validated by the broader community, leading in part to the development of the peer-review process and need for scientific-training credentials, which in turn further contributed to the distinction between amateur and scientist. In this system, a scientific community became "entrusted" with vetting which people fulfilled the specific expectations of a professional and which new knowledge passed muster. Trust became a type of currency that was only handed out sparingly to the knowledge that met the community standards.

Where does this put us with games? In them, not only are we removing the person from the knowledge, but we are again changing the system that we use to vet that knowledge. We are putting the knowledge creation in the hands of not just one person or research team, but a collective of players and entrusting them with it. And we are not only entrusting multiple people but, often, many anonymous people, who are not only *not* scientists but proud amateurs. How can we trust that the knowledge game players are truly contributing to knowledge if they do not have the training, the expertise, the authority, the rigorous peer review, and the other culturally prescribed validations that we, as a society, require?

Can we really get accurate, realistic results, and solutions from games? Can we trust the knowledge created by a game if, despite a game's purposeful goals, some people are playing the game simply "for fun" while others are diligently trying to solve scientific or social scientific puzzles? In *Beyond*

Pleasure and Pain, E. Tory Higgins describes a study using a paired-associate learning task. Some participants were told to have fun while learning the necessary material, while other participants were told that the task would be tedious and serious. Participants were also divided by high and low importance: some received instructions that the task was of high importance and others got instructions that it was just a pilot study. Higgins found that those people in the "high importance and high fun" group actually had a lower performance than those in the "high importance and low fun" one.[42] The result is counterintuitive because we expect fun to be motivating and enhance our performance. This is typically true if we presume that the activity will be fun, such as a game. The researchers explain, however, that since people usually do not expect that learning will be fun, this ends up actually hurting performance on a learning-related task. What then would happen with a knowledge game, which could be framed as both fun *and* serious? The way a knowledge game is framed—but also our expectations for fun experiences such as games and serious activities such as knowledge production—may affect not only how it is perceived, trusted, or accepted but also our performance in it. More research is necessary to parse out whether we should trust the accuracy of knowledge games, particularly if they are framed as being fun.

As we think about our conceptions of amateur and professional, leisure and work, player and scientist, fun and seriousness—all of which knowledge games call into question—we also need to rethink the entire way we create and vet knowledge as a society.

It is no surprise that this questioning of knowledge creation is coming at a time of crisis in the peer-review process, changing data-analysis methodologies, shifting resources for research, and the transformation of other markers of the scientific and knowledge communities.[43] In chapter 8, I further question trust in knowledge, by looking at how we analyze and process data and use it to vet new contributions and establish new knowledge.

The Social in Science

I have discussed the role of amateurs in scientific knowledge production, but what about not-so-scientific problems, such as bullying or school dropout prevention, or problems that mix scientific and social aspects, like depression, global warming, or suffrage? Science is different, right? It's all facts and figures, which can be potentially vetted by the scientific community, given the correct training, expertise, and ability. Science is not sullied by pesky emotions, biases, or political maneuverings. However, as I discussed earlier, scientific practice was influenced by, and influences, communicative technologies and culture, whether the printing press or Internet. I have also considered how the process of scientific knowledge creation has been socially, economically, and culturally constructed—and prone to errors and inaccuracies. It is a system to which we subscribe, but it is not the only type of system that could exist.

So is science really so different from the social? As Mark Brown observes, "There is nothing especially scientific about the creation of scientific knowledge." We popularly conceive of science as exact, precise, and technical, and social science as more nuanced, fuzzy, and human. In fact, these conceptions help to further reify the scientific profession and separate it from amateur and public activity. When, at a conference, I mentioned the concept of using knowledge games for social and psychological purposes in addition to scientific ones, some audience members were troubled. In particular, the possible personal privacy issues bothered them—issues that are not as regularly considered as problematic for citizen science or crowdsourcing in general, as science is seen as separate from the personal.

Yet the process of scientific knowledge creation is much more like that of other pursuits—sticky, tricky, and uncertain. Science is conducted by human beings, as fallible and apt to err as any of us. Whether we like it or not, we are part of a sociocultural system that determines the norms of what it means to be accurate and scientific, and what is even worthy of being researched. Explains Brown, citing Latour, "to conceive of humanity and technology as polar opposites is, in effect, to wish away humanity: we are sociotechnical animals, and each human interaction is sociotechnical . . . There is no such thing as an original, nonmaterial, nontechnical human who imposes its will on inert and shapeless matter. Human lives are permeated by technical objects and these objects play an important role in constituting human subjectivity."[44] So while we often popularly conceive of science as separate and outside of culture, politics, or social structures—hovering above all that human stuff—I make the assumption throughout this book that science is not insulated from culture and society.[45]

This means that we need to analyze the full implications of knowledge games—whether for scientific knowledge production or beyond. Privacy, ethics, and other issues should be deliberated for those using, designing, and playing all types of knowledge games, citizen science, and crowdsourcing activities. That said, particular consideration to the (mis)perception of the insularity of science from culture needs to be made when analyzing the potentials and challenges of knowledge games.[46]

Moreover, amateur participants are already intimately involved in contributing to not-solely-scientific-focused projects. People are annotating and tagging diaries in Operation War Diary to learn more about World War II, such as the military strategies enacted during that period. They are playing language games such as *VerbCorner* and *Which English?* to share how they use language in different contexts. People can even look privacy right in its face and submit reports of their sexual activity on the Kinsey Reporter app, or grant access to the Immersion Project to collect metadata on one's own Gmail account and generate maps of personal contacts and interactions.[47]

In the next chapter, I look closer at participation, including who participates and who has access to participation. What are the ways we can conceptualize participation in and around knowledge games, and how can we further unpack the complex relationship among games, work, fun, and labor?

CHAPTER SEVEN

Participation

What color is that moth wing? Does it have stripes, spots, or colored tips? In the game *Happy Moths*, you can answer these questions, participate in a real-world citizen science project, and contribute to our knowledge of insects. *Happy Moths* is part of a suite of games called *Citizen Sort*, a research initiative by Syracuse University's School of Information Studies to further our knowledge of biology and better understand how to motivate more people to participate in citizen science projects.[1]

Likewise, in the alternate reality game (ARG) *Evoke*, and its first module, *Urgent Evoke*, which ran for ten weeks in 2011, anyone could participate in providing innovative solutions to pressing problems, such as hunger, poverty, or human rights issues. Each week, participants would get a new problem to solve, which was introduced in a comic book style. Participants could learn about and research the problems by going on a mission or quest, such as volunteering in one's community (e.g., providing meals for an elderly or homebound person), and then documenting it with a video or blog post to earn points. Jane McGonigal, one of the game's designers, explains that *Evoke* was designed to "empower young people all over the world, especially in Africa, to start actively tackling the world's most urgent problems—poverty, hunger, sustainable energy, clean-water access, natural disaster preparation, human rights."[2]

While both projects have clear, valuable goals, *Happy Moths* and *Urgent Evoke* feel less like games and more like repetitive exercises. *Happy Moths* uses a streamlined interface and invites clear interactions, but the game does not change from round to round. Instead, you spend the entire game as a taxonomist, hoping that each moth you classify is correct. Likewise, *Urgent Evoke* is unique and ambitious, particularly in motivating a large number of participants to take knowledge and skills back to their real-world communities. However, in practice, *Urgent Evoke* seemed more focused on crowdsourcing ideas, player-created media, and participant testimonies, as well as gamifying volunteer and community work, rather than really supporting the collaborative connections and novel interactions that games could afford, connections that might lead to problem solving and real-world

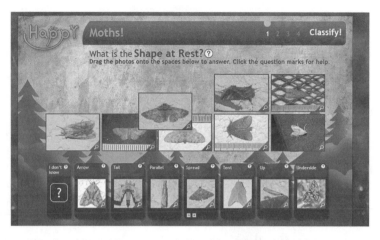

In *Happy Moths*, players categorize and classify moths based on a variety of dimensions, including color and wing shape. Source: Nathan Prestopnik, Kevin Crowston, and Jun Wang, "Exploring Data Quality in Games with a Purpose," in *iConference 2014 Proceedings*, doi:10.9776/14066. (This work was partially supported by the US NSF under grant SOCS 09-68470.)

change. Moreover, since the majority of the participants in *Urgent Evoke* were already involved in nongovernmental and activist organizations, the game mainly reached participants already engaged in these kinds of actions in the real world.[3]

Each game brings up questions of participation. Who is participating and what is the tenor of the participation? Knowledge games are not successful if there are not enough engaged participants. But beyond that, how are people participating in these games? Are *Urgent Evoke* and *Happy Moths* truly games—or do they venture too far into the realm of work? Are they fun enough?

In chapter 1, I discussed how people can contribute to scientific, humanistic, and social scientific knowledge. In chapter 6, I began to question the historical relationships among fun, work, leisure, and play. But we did not delve into models of how people actually participate in systems that enable knowledge production and how this participation might relate to work, labor, or play. What are the ways people are allowed (or not allowed) to take part in games, in work, and in knowledge production? In what ways can we learn about society through the societies created around and within games? Is all participation created equally? What happens when previously distinct conceptions of participation begin to blur?

How People Participate

When it comes to participation, are all contributions and contributors created equal? How do people participate in efforts such as crowdsourcing activities or knowledge games? Zhao and Zhu argue that understanding the quantity and effort of contributions is useful, as is understanding the process of contribution, including "what happens when crowds select tasks, compete or collaborate with others, and submit feedbacks." Raddick et al. observed that people may participate in Galaxy Zoo very differently and engagement can range from deep to superficial. "Some classify galaxies only once, some return to classify thousands of galaxies, and some go far beyond the basic task of classifying galaxies into a range of different involvements with the project, such as reading the Galaxy Zoo blog or engaging in deeper scientific research on new astronomical objects in the Galaxy Zoo forum."[4]

Raddick et al. acknowledge that participation in Galaxy Zoo varies, but they do not segment the participants in their study by how often they contribute or the quality of the galaxy categorizing contributions. Yet this type of information would have helped us better understand the landscape of what motivates different participants, and how to better design for them. For instance, who are the supercontributors? Do they have different needs than the dabblers? Do different design elements motivate and sustain them? If you are designing a knowledge game, do you want to spend resources to get as many people as possible, hoping that some are supercontributors, or do you want to target only the supercontributors?

Stewart, Lubensky, and Huerta studied the mechanism of crowdsourcing in a challenge they devised. They found that participation in online communities is typically imbalanced and follows the 90-9-1 rule, in which "(a) 90% of users are 'lurkers' (i.e., they read or observe but don't contribute), (b) 9% of users contribute from time to time, but other priorities dominate their time, (c) 1% of users participate very often and account for most contributions." They use this rule to propose the SCOUT model, labeling the aforementioned three groups the outlier (OUT) (90%), contributor (C) (9%), and supercontributor (S) (1%) groups. You can think of this model in terms of your friends on Facebook, Instagram, or Twitter, where about 1% post constantly (or so it seems), 9% post pretty often, and the majority may click "Like" or view posts but only rarely actually post themselves. This is not necessarily a problem for every platform because we need listeners and viewers, and not just constant content creators. In a crowdsourcing plat-

Galaxy Zoo has a few "spinoffs." In Radio Galaxy Zoo, participants can annotate images of potentially erupting black holes, which could be dragging jets of material into them. Source: Zooniverse

form or knowledge game, however, we may want to motivate as many people to participate as deeply as possible. Stewart, Lubensky, and Huerta argue that crowdsourcing platform creators can encourage further participation so that only 33% (rather than 90%) are outliers and 66% (rather than 9%) are moderate contributors. To do this, they tweaked the reward structure on their platform by providing a multi-tiered incentive model, which did achieve their desired change in moderate contribution (though they did not change the behavior of supercontributors).[5] The researchers, however, did not test whether this model could be applied to other crowdsourcing contexts.

Even if we understood why each person participated on a crowdsourcing platform or in a knowledge game—and we were able to segment the audience and provide appropriate incentives or experiences so that people were more likely to contribute—do we want or need all of these participants? While you may want lots of people participating intensely, you also want their contributions to be accurate, appropriate, and engaged. Is there a baseline level of participation that is necessary for success? Is it beneficial to have large quantities of people participating, or is quality of participation more important? And do different factors support each?

Nov, Arazy, and Anderson investigated three citizen science projects, focusing primarily on Stardust@home, and looked at the motivational factors

driving contribution to the project. Applying a model from social movement participation, they found that different factors determined contribution quantity and quality. The results suggested that the citizen science participants were motivated to contribute more (*quantity*) by what they called
collective motives (those motives associated with the project's collective
goals), norm-oriented motives (motives related to how family and friends
might be expected to react), reputation (a type of extrinsic motivation), and
intrinsic motives. The *quality*, or degree of excellence of participation, was
only affected by two factors: collective motives and reputation.[6] Thus, different factors may affect people wanting to participation more, rather than
being moved to provide higher quality contributions. These relationships
should be considered further in relation to knowledge games specifically.

The Participation Gap

In addition to getting enough high-quality or appropriate participation, designers and producers of knowledge games may also want to ensure that the
right people are participating. Who should actually be contributing through
these types of games?

Today, only a narrow proportion of the contributors online contribute to
citizen science project or knowledge games. While there are 300,000 registered *Foldit* players (2,000 are active) and 850,000 participants in websites
such as Test My Brain, a site that features online psychological tests that
people can take, more than half of all Americans (approximately 188 million) play video games.[7]

In fact, most platforms are imbalanced in terms of their makeup. Neil
Seeman notes that social media, such as Instagram, skew young, and platforms are often discussed in relation to their primary audience (such as
Reddit's young male demographic or Pinterest's middle-aged females and
moms). Even the primary language of the Internet (English) can bias who
participates and any results from research conducted on those platforms.
Where we participate online and whom we participate with are affected by
and affect our actual participation. What types of people we attract on a
platform automatically biases our population, and thereby any results we
might claim to obtain through researching it. For example, mining Twitter,
Facebook or other social media platforms to better understand people's ethical behavior could render an irrelevant result, particularly if we wanted
to include those who rarely use these platforms. If we poll our Facebook
friends, the result represents our social network's behaviors and attitudes,

not necessarily that of the world at large. There is always the risk in any given platform that the opinions of the few but loud (such as those with bigger, Thunderclap-charged voices) could become accepted over that of the many or the quieter. The potential ease of using social media or a game to recruit or sustain participants should not replace the process of finding the right participants or seeking diverse views and opinions.[8]

What about volunteers in general? Those who volunteer in general typically are well educated, have higher income, and are more religious. Wilson suggests that the relationship between educational achievement and higher volunteering levels is due to belonging to more organizations but also to having "broader horizons, as measured by attention to current affairs, higher levels of cognitive competence, and higher status jobs." Building on Shye's findings, Wilson proposes that workers who are more satisfied with their jobs and feel rewarded by them may be more likely to volunteer. And while higher income has an effect on volunteer participation, it is "not linear, volunteering being most popular among middle-income households."[9]

However, it is sometimes difficult to unravel the complexities of who volunteers and why. For example, women may participate in volunteer work but not identify it as such. Women's participation in volunteer work is often taken for granted, whereas a man might do the same type of work and it would be considered rare. Wilson gives the example of a breast cancer campaign, where female participants were seen as a given while any male participant performing the same tasks was seen as a hero.[10]

Crowdsourcing platforms are also demographically skewed. As noted on page 81, the participants in Brabham's iStockphoto study skewed male, white, married, and middle or upper class. Lakhani et al. found that the InnoCentive platform's demographics skewed white and middle aged, with a majority holding a doctorate. This suggests that only a small subset of the population is participating, creating a "contribution divide" between those who contribute and those who do not.[11]

Gaming further complicates these demographic tendencies. To understand potential participation in knowledge games, we need to consider the different rates of participation in gaming in general and the types of games that may attract certain groups of people. In 2015, the average age of a game player was 35, so if we want to target older participants to engage in knowledge production, gaming may not be the way to go. Although 56% of game players are male and 44% female, we cannot safely conclude that the same proportion of participation by gender would occur in knowledge games.[12]

On the other hand, general statistics about game use may not be applicable to knowledge games, particularly as we think back to the research in chapter 4 about youth, motivation, and citizen science games, and the finding that many avid game players may just not want to play these games.

Beyond considering who participates in games or other platforms, we also need to consider the structural issues that encourage or discourage participation. Not everyone has access to Internet-enabled technologies, such as computers with home Web access. For example, only some people have access to the equipment necessary to participate in a particular citizen science project or knowledge game, such as the special type of drone, camera, or mobile device needed to play the *Astro Drone Crowdsourcing Game*.[13] Only those with an iPhone or iPad can play *Reverse the Odds*.

People also need access to time to play a knowledge game or engage in a crowdsourcing task. Some people can do this because they are able to feed and shelter their family with a job that pays enough and allows time for leisure. Some have spouses or parents who help care for the family, bills, and housework so that they can also spend time on other pursuits.

Beyond having the right equipment or enough time to offer, there are other differences that enable or disable participation, such as technical, literacy, and production skill abilities. Knowledge game participants may need to be able to use and apply information literacy skills or access and employ gaming skills to be effective participants in a game.[14] Non-English speakers may not be able to comprehend the necessary instructions to play games that focus on English language usage such as *VerbCorner* or *Which English?* A toddler may not yet have the reading, coordination, or gaming ability to play *EteRNA* or *Foldit*. Differences in skill, experience, and literacy levels among potential players will affect their participation. People cannot play, solve problems, or participate without the necessary basic skill levels or literacies, which also limits the reach of knowledge games.

Participators, Unite!

What are the other ways that participation is negotiated structurally and societally? How privy are we to participating in various modes of production in our society—from media production to knowledge creation? As we saw in the last chapter, amateur participation in professional knowledge-making fields, such as scientific knowledge production, is nothing new. Likewise, consumer participation in the creative fields—such as television, magazines, newspapers, and film—has also taken place historically, even if it has not been as immediately available and accessible as it is now.[15] Are we moving toward more open and equitable participation in all of these realms? Have the access points for amateurs to produce knowledge expanded?

Axel Bruns describes how people are now readily participating in the creative processes around entertainment and news, such as in the development of open-source software, blogging news events, or creating videos and posting them to YouTube. Bruns calls this type of activity "produsage" or the "simultaneous production and usage," meaning that people are acting as both consumers and creators of content and experiences, disrupting the typical top-down hierarchy of production.[16]

Produsage is similar to an older term, "prosumer," or a consumer who practices professional activities, which could encompass anything from assembling IKEA furniture to pumping one's own gas. The terms "co-design" and "co-creation" are similar too. According to Banks and Potts, co-creation and a co-creative culture occur when consumers are actively involved in supporting some aspect of a product's design, marketing, distribution, or development.[17]

Players have found many ways to participate more deeply in the co-design or co-creation of the game experience, either through vetted (e.g., by the game developers) or backdoor means (e.g., without explicit support by the developers). For example, participants might mod, or modify, an existing game. Modding is the act of changing hardware, software, or any other object or experience in a way that was not originally intended by the designer or customizing the game to one's own specifications and preferences. For instance, players of *Minecraft* can add new content, such as a set of tools for building complex railways within the game, which other users can also use and which adds to the overall content of the game.[18] While not considered modding per se, participants can contribute content and build everything from full-scale models of the Taj Mahal to fictional towns and villages within the *Minecraft* environment.

Likewise, players of *Spore* can create new monsterlike avatars using the developer-vetted Creature Editor, which also reduces the number of characters the developers have to create themselves. Within 18 days of the release of the *Spore* Creature Editor on June 18, 2008, participants had already designed 1,589,000 species, or the number of known species on Earth.[19] Players can also create and even sell items such as weapons, hats, or maps in *Team Fortress 2* through a Valve-created contribution site. The most sought-after items are then officially added to the game, and content creators from the game community can even earn money for their created artifacts.

Thus, game players may have many ways they can act as "produsers"—or, perhaps, "prodplayers." Could knowledge game players also be seen as prod-playing, as they are playing, using, and creating new knowledge alongside the game's designers? Could we conceive of knowledge games as a place where amateurs might co-research with professionals?

Participation Models

Beyond Brun's "produsage," how else can we conceive of the types of participation knowledge games afford? First, we can look at crowdsourcing and citizen science participation models. Shirk et al. devised five different models of participation by the public in citizen science projects.

1. *Contractual*: These relationships occur when communities ask professionals to conduct specific research and report on it.
2. *Contributory*: People from the public contribute by collecting data but do not necessarily interact with each other. The citizens are gatherers and laborers, bringing the data back to the scientists. This type of project is usually designed by scientists.
3. *Collaborative*: These are projects designed by scientists where members of the public directly work with the scientists to analyze and interpret the data. Citizens might also help design the project and disseminate its results.
4. *Co-created*: People are more actively involved in all stages, including designing the project itself and/or interpreting any data. Members of the public and scientists might design and initiate the project together.
5. *Collegial*: Members of the public or nonprofessionals conduct independent research, with varying expectations on how much of it will actually be accepted or recognized by the scientific community.[20]

We can also look at models of more general public participation. In *Wealth of Networks*, Yochai Benkler describes a commons-based peer-production model. He argues that knowledge and culture are being created more and more outside of the typical institutions, markets, and firms, through new nonmarket modes of production. The impetus for this new model stems in part from newly available and accessible communication platforms (such as the Web, online games, and video-editing apps), which enable nonprofessionals to participate in more sophisticated hands-on production. Through the commons-based peer-production model, Benkler imagines a more democratic distribution of computing power or creative tools, as well as training on those tools, and the greater ability to share and distribute one's products with others.[21]

Likewise, Lawrence Lessig argues that there should be open access to, use of, and facility with intellectual property. "With respect to intellectual property, I argue against code that tracks reading and in favor of code that guarantees a large space for an intellectual commons." With more open access to using, reshaping, and reconfiguring this so-called "property," we can more easily attain a commons-based peer-production model, which can enable more democratic development and distribution of creative goods.[22]

Jenkins et al. describe "participatory cultures" in which people are active participants in media creation, reconfiguration, revision, and reconstitution. In participatory cultures, people come together to collectively participate in story worlds where they can alter story lines, create new characters, provide new subplots, or spoil reality shows. The authors portray participatory cultures as low-barrier-to-entry cultures with opportunities for artistic expression and civic engagement, social connectedness, mentorship relationships among participants, and the ability to meaningfully create and share creations. Examples of participatory cultures include affiliations (Facebook groups, game player communities), expressions (fan-fiction writing, modding), collaborative problem solving (*Wikipedia*, ARGs), and circulations (blogging and podcasting).[23]

Another way to conceive of the affiliations in and around knowledge games is as what Jean Lave and Etienne Wenger call a community of practice. In a community of practice, people come together around a specific activity or practice.[24] These communities can even serve to legitimize and approve the practice of spending time and effort on a particular task, such as folding proteins in *Foldit* or creating hats in *Team Fortress 2*. A community of practice may be a useful label for a group of Threadless artists who share artistic expertise or a group on Galaxy Zoo sharing tips on how to better classify galaxies. A defining aspect of such groups is how people are able to learn through the process of "legitimate peripheral participation," a process that enables novices to embody the practices of that community by working on peripheral activities and tasks, and over time, taking on more responsibilities central to the organization. Paraphrasing Lave and Wenger, Zagal and Bruckman write, "An extended period of legitimate peripherality provides novices with opportunities to make the culture of practice their own."[25]

Gee and Hayes critique the idea of a community of practice, arguing that it works better as a model for describing face-to-face interactions. Instead, they use the term "affinity spaces," which they feel more readily describes communities that form through technologically supported platforms that function across time and distance. Affinity spaces are self-organizing communities that emerge from voluntary interactions and the sharing of knowledge, experience, and interest. But they are not the same as communities, for communities are deeper, stronger, and more intimate than affinity spaces. Some affinity spaces take place wholly offline, some wholly online, and some provide people with a feeling of belonging, whereas others are more exclusive. Affinity spaces can be potentially found within and around games as well, such as in the chat rooms of *EyeWire* or the forums of *EteRNA*. Debate, deliberation, knowledge sharing, apprenticeship learning, and collaborative problem solving sometimes occur within these affinity spaces, as well as within a game itself, depending on how the game and the surrounding communities are designed and emerge. Gee and Hayes looked at *The Sims* and its affinity spaces and derived 15 features of them, including that the participants have common interests, goals, and shared passions; that novices and those who are more experienced tend to share spaces; and that participants can produce content and knowledge, not just consume it, if they want to.[26]

In affinity spaces, prior knowledge and expertise become shared, and people teach each other both specialist and more general knowledge. Newcomers to *EyeWire* or *EteRNA* can easily get an introduction to the gameplay, while more experienced players can share specific problem-solving tips or techniques. In addition, affinity spaces encourage the cultivation of both individual knowledge (solely within one person) and distributed knowledge (the total knowledge found across each person, object, or site involved in the affinity space). As discussed in chapter 5, a group's ability to solve problems or meet goals can become greater than the sum of its parts. In *EteRNA* or *EyeWire*, a person might pose a question to a forum or in-game chat and then use the expertise of the group to find a solution, rather than simply relying on his or her individual knowledge. Moreover, experts and learners continually switch roles and take on the tasks of both teaching and learning. Experts and learners give back to each other and are reciprocally engaged.[27]

Finally, Gee and Hayes note that affinity spaces support the use of both tacit and explicit knowledge.[28] Tacit knowledge is knowledge that people might have but may not be able to explicitly articulate, such as understanding how to play a game and avoid certain types of obstacles or knowing how to care for your baby by observing and practicing with others in your community rather than by reading manuals. Explicit knowledge is knowledge that is directly expressed, such as a teacher indicating the capital of a country or a parent defining a word for a child rather than the child just picking up its meaning from contextual clues. The incorporation of both tacit and explicit types of knowledge shared through an affinity group may more readily help participants of a knowledge game learn strategies that even an in-game tutorial or set of instructions cannot teach them.

Although these three concepts of participation (participatory culture, affinity spaces, and communities of practice) are slightly different, each conveys the notion that there is some type of participatory knowledge node, which is shared, negotiated, and contributed to by a group of people. This knowledge node could be supported by information systems and technological platforms, though the form, culture, and interactions of these groups predate and do not rely on the advent of new technologies.

An overriding theme with these concepts of participation is that participation is not just about the technology or the platform used; rather, it is about the culture and context and all of the practices, institutions, and values embedded in and around them. The ability of an online or virtual community to enable participation relies not just on a platform's technological specifications but also on a complex mix of social, psychological, and cultural elements, which can interact with physical and technological constraints.[29] This is similar, to, say, how any human communities function and evolve in relation to the physical resources available, organizational structures, values, and interactions with other communities.

Work Models

In general, Benkler, Jenkins, and Gee are mostly optimistic about group interactions and view participation as achievable, edifying, and effective. They seem to suggest that learning, idea exchange, creative products, and knowledge making can come from these visions of participation—and that they are beneficial forces. Benkler and Jenkins also appear to suggest that public participation is generally advantageous, both to the user and to the original creators or corporate interests of a platform or game. For example, fan participation in a transmedia world for kids such as The 39 Clues is beneficial to both readers and Scholastic (the owner), as it further engages the users, provides them with opportunities to contribute to the overall story-world, and allows them to practice media literacy, media creation, and fluency skills.[30] It can also enhance profits for Scholastic. While Benkler and Jenkins might contend that modding or fan fiction are valuable to all, some researchers have questioned whether this type of participation is actually more problematic. Some researchers consider fan contribution and other forms of participation, such as through crowdsourcing or even citizen science, as more akin to exploitative labor. This leads to further questions. When does play become work and how does work tie into play? When does participation in these platforms become inappropriate or even exploitative? What is the role of fun in all of this?

First we need to distinguish what is labor and what is work. Marx, in *Capital*, maintains that labor is performative, a process of discovering our passions and interests, as well as our overall self-concept.[31] There are different types of labor, such as traditional labor, direct labor, immaterial labor, and indirect labor. These types are important to differentiate, particularly in the emerging digital economy, or the "new economy based on the networking of human intelligence." Unlike traditional labor, such as the labor performed by factory or store workers, how do we organize and manage the work of the digital economy and "knowledge" labor? Whereas old media has clear roles for production (e.g., writers, producers, performers), the so-called new media has less clear roles. Tiziana Terranova argues that the digital economy is about "specific forms of production (Web design, multimedia production . . .), but is also about forms of labor we do not immediately recognize as such: chat, real-life stories, mailing lists, amateur newsletters, and so on."[32]

According to Terranova the new digital economy includes both direct labor and immaterial labor. Direct labor involves skills such as computer programming or Web-content writing, and these laborers might create defined, bounded products, such as software, data, and music. Immaterial labor, as Terranova calls it, involves more abstract and intangible forms of work, such as providing opinions, commenting on fashions, or defining standards of quality.[33] For instance, in the new digital economy, immaterial labor might include posting articles on Facebook, commenting on Fashion Week trends, creating a Pinterest board of weekly planner inspirations, or e-mailing a recipe to a friend.[34] New digital economy laborers could include knowledge game players, who might perform activities such as inputting data, analyzing trends, submitting opinions, solving puzzles, or reviewing information. Should this be categorized as immaterial labor?

Moreover, the work of the digital economy never ends, quite literally. Terranova contends that "the Internet is about the extraction of value out of continuous, updateable work, and it is extremely labor intensive."[35] For example, as soon as a website (1.0) is launched, the team that worked on it is already working on its relaunch (2.0), and making constant content revisions, image swaps, and phasing in new features (with each new version designated as 1.x). The work of the website is never done, but it is not necessarily defined as direct labor.

Christian Fuchs identifies other types of laborers, such as the people who do the work of society and are still working from within the system of capitalism, wages, and labor, though not necessarily getting paid for any of it. These indirect laborers "produce and reproduce the social conditions of the existence of capital and wage labour such as education, social relationships, affects, communication, sex, housework, common knowledge in everyday life, natural resources, nurture, care, etc. These are forms of unpaid labour that are necessary for the existence of society."[36] For example, my own care of my children is unpaid work, and so is the time my husband spends washing dishes or doing laundry, but these activities are necessary for maintaining society, as well as to ensure our continued ability to do paid labor and contribute to capital. Likewise, Postigo calls this type of labor "passionate labor" because it provides emotional and spiritual return, rather than financial incentives.[37] (Although we may wonder how passionate we feel about washing dirty dishes or folding laundry.)

Direct labor also seems to be undervalued in the digital economy, even if it takes a lot of time, effort, and skill, such as the through the time spent programming online software or creating digital artifacts for a website. For example, when I worked at an educational technology company, we conducted a proprietary analysis on how much money consumers were willing to pay for tangible (boxed) educational software versus the same materials provided online. Although direct consumers were willing to pay for the software that came in an actual box, at the time (2009) no one in the study was willing to pay for resources provided online (for example, through a subscription service or one-time fee for downloading the materials). While the hesitancy to pay for digital goods has allayed in the last few years, consumers, producers, and owners may still not be able to visualize as easily the value of the direct labor involved in the creation of digital products.

Another concept useful to our understanding of participation in knowledge games is the notion of the informal economy. The informal economy consists of many forms of amateur creation—from pirated music to zines, from secret game mods to machinima.[38] Some of this labor is provided for free or unpaid, though some of the activity earns money in alternative markets (e.g., black markets, gray markets, or second-hand markets). The informal economy has moved from being something behind the scenes and needing to be squelched and regulated by the owners of production to becoming a sought-after component of the formal economy.[39] The activities of the informal economy are still work that has value and takes time, no matter how it is integrated into traditional modes of production.

The Internet, argues Terranova, relies on free labor and amateur production to sustain it, including "labor" such as e-mailing, IMing, website updating, or contributing comments, or even Threadless T-shirt submissions. Users of a site keep it alive and sustain it, whether via Facebook posts or Quora questions. Watching, reading, customizing avatars, creating Spotify playlists, *Spore* creatures, and swiping right on Tinder are all work. As Grossman explains, "We didn't just watch, we also worked. Like crazy. We made Facebook profiles and Second Life avatars and reviewed books at Amazon and recorded podcasts."[40]

Games can also rely on this type of labor. As discussed earlier, modding and user-generated content in games can function as free labor that a game company can use to reduce costs. *Spore* developer Will Wright has commented that the content creation afforded by the Creature Editor was due in part to the increasing costs of developing a game, such as having to create so many different art assets.[41] Enabling user-generated "monsters" with *Spore*'s content creator gave players the so-called freedom to design their own characters, but in essence it also was about outsourcing labor to the public. And it was successful, generating millions of free art elements for the game, as noted above. Likewise, we could conceive of all crowdsourcing and citizen science activities as being outsourced labor to the public. Categorizing galaxies on Galaxy Zoo, sorting moths in *Happy Moths*, and folding proteins in *Foldit* could all be considered free labor that the public does for "the good of science," which also defrays the costs of paying scientists or other workers.

But is this type of free labor in games or beyond really all that new, or so bad?

The Audience as Laborer

Well before the advent of the Internet and other connected technologies, the activities that are performed by an audience member or "user" of a mediated experience had been cited as labor or work. Hector Postigo points to the idea of the "working audience," building on a concept from Sut Jhally. Likewise, Dallas Smythe argues that the audience is a commodity that is created, bought, and sold. "Because audience power is produced, sold, purchased and consumed, it commands a price and is a commodity . . . You audience members contribute your unpaid work time and in exchange you receive the program material and the explicit advertisements." Although the audience is unpaid, it is, and has always been, a commodity that is bought and sold in the capital system. Audience work is both a product and part of the system that propagates capital and labor relationships. Fuchs argues that the work that the user does, whether looking at videos online, performing Google searches, contributing data, providing opinions of T-shirts, or posting updates on Facebook, makes up an "audience commodity that is sold to advertisers."[42]

Fuchs makes the distinction between the audience commodity that derives from mass media, such as television, film, news, and radio, and that which emerges from newer media. He argues that through Internet-enabled platforms, we are more able to participate as content producers and co-creators or take part in a community of media creation, which conflates these relationships.[43] Thus, the connection between user and producer, and between audience and owner, gets further complicated in the new digital economy. Whereas the distinction between television advertiser and television watcher was once clearer, the relationship between user and creator is now more clouded, whether the labor is done as part of a social network or involves tweeting a news article, contributing to a blog or a recipe database, or pinning media on Pinterest.

Complicating things further, some of these audience "laborers" may learn more about their passions through these activities or find new careers through these initially avocational activities, such as the mompreneur who creates fun activities for her kids, blogs about it, and turns it into a profitable subscription-based "activity kits" business. Some users may discover new talents or interests, such as in writing or photography, and even get paid for it on Elance or iStockphoto. Some may find that they enjoy television shows or books more through writing and sharing fan fiction or helping to spoil *The Bachelor*; in very rare cases, a writer can turn her fan-fiction novel into the bestselling *Fifty Shades of Grey* or a blogger can turn his spoiler site into the lucrative Reality Steve. Are these productive creative activities—which could have benefits for both the audience and the corporate entities—really just another form of work? If so, does it matter?

All Work and No Fun

Can we also apply the "audience as laborer" concept to knowledge games and their players? Is playing a game a form of audience work? Is play always work? As described in chapter 6, we frame games as, by definition, fun and separate from work. But as Bogost relates, perhaps fun is "excreted" from the structure of gameplay, its wretchedness, and even from its work. When is play just play, and when is it a form of labor? When is the work we do in a game exploitative, and when does it render a game no longer a game?

To unpack these questions, we need to first understand the connection between work and games. How is playing a game like work? In 2005, Julian Kücklich introduced the term "playbour" (a portmanteau of play and labour), suggesting the relationship between play and productivity, work and labor.[44] Likewise, Hector Postigo argues that play is part of the production process, such that in the making of gameplay, play and production become unified and "the features that frame gameplay on YouTube extend beyond the platform's architecture to a matrix of technologies that work with it, by design or as a matter of coincidence, to allow (1) gameplay, (2) making gameplay, and (3) making game pay."[45]

How does this translate to knowledge games? In knowledge games, players are playing *and* their play directly leads to productive purposes that are potentially valuable to society, such as knowledge making and problem solving. If this type of play is labor, can it still be fun, personally meaningful, and intrinsically enriching? Can it be considered leisure? In one sense, the concept that all play is productive and that all players are laborers, rather than simply frivolous leisure seekers, helps legitimize the position of knowledge games as games. In other words, all game players are always laboring, so knowledge game players are no different and no more problematic. At the same time, a lack of clear boundaries between play and labor has unintended, and perhaps undesirable, consequences. For example, because of the increase in crowdsourcing and play labor, the working day is creeping into our leisure time, maintains Fuchs. Using social media creation as a model, he argues that capitalism sees an opening (which is our leisure time) and seeps into it as a way to expand labor time and generate profit beyond the typical working hours. Fuchs might claim that similarly, knowledge gaming could convert "non-commodified leisure time into productive labour time that generates value and profit for capital." In other words, converting leisure into labor is just another way for the tentacles of capitalism to infiltrate more deeply, even if it is supposedly time spent on personal or creative pursuits, the products of capitalism, or nonwork time. Thus, leisure time creates extra or surplus value produced for capitalism, particularly in what is generated through advertising and surveillance (see chapter 8) but also by the knowledge produced by knowledge game players.[46] In fact, surplus value is created for companies such as Google and Facebook not just by those paid workers who are programming or doing interface design, marketing, or other direct wage-labor activities but also by those participants who are producing free user-generated content, such as posts, comments, and likes, or even websites, blogs, and mods.

Ironically, therefore, at the moment when we may have more disposable time and more "leisure" time to, for example, play games, capitalism is turning that time into productive and surplus value, according to Fuchs. We can argue that uses of our disposable time, such as knowledge gaming, game playing, and crowdsourcing are therefore being commodified by capital.[47]

Knowledge gaming labor is provided for free, but is it provided willingly? As Tiziana Terranova writes, "Free labor, however, is not necessarily exploited labor."[48] Is it problematic that knowledge gaming is actually free labor and work yet also provides value for capital?

Houston, Is There a Problem?

People may spend a significant portion of their leisure time on activities such as crowdsourcing, citizen science, and knowledge games. At what point are players and participants giving too much? Under what conditions could we consider knowledge games exploitative?

Usually when we ask questions of video games and exploitation, we consider issues such as violence, sexism, addiction, or racism, or we examine the free-to-play or other business models and their potential exploitation of players and cash "cows." In terms of game production, we might also consider the long, sometimes unpaid, work hours and crunch-time overnights that in-house game developers, artists, writers, designers, quality assurance testers, and managers may contribute to a game's development.[49] Or we might consider how game companies outsource work to laborers in developing countries and try to take advantage of lower wages and less-regulated work conditions.

But in this book, we are questioning whether playing a game—even if it is entirely free of charge and potentially free of long in-house or outsourced developer hours—could be considered exploitative. Moreover, if knowledge game playing is voluntary, can we ever call it exploitative? If there are benefits to game playing, including honing creative skills; media fluency; learning about science, humanities, or social science; feeling good about helping out an organization or cause; community membership; or gaining leadership skills, are there instances where this type of game labor could be exploitative? Does it exploit the player even if the outcomes of play are edifying for society? How do we resolve the complications that this play might have the potential power to contribute significantly to society and the common good?

Marx contends that exploitation is a basic function of capitalism, as the capitalist production's primary goal is to exploit labor as much as possible.[50] Nancy Holmstrom maintains that exploitation occurs when producers of a product do not control it and are alienated from it. Mark Andrejevic suggests that a key aspect of exploitative activities is that our work is transformed unrecognizably and serves needs and desires that are separate from our own.[51]

Fuchs and Terranova argue that the same hegemonic structures that inform capitalism also affect social, cultural, and economic relationships in the digital economy. Far from being a more democratic, antihierarchical construction, the digital economy may actually replicate those structures and perhaps even may be more exploitative because they are more hidden, unpaid, pervade leisure time, and are couched as "fun." Thus, Fuchs contends, crowdsourcing is more about exploitation than democratization and that it reaffirms capitalism. These relationships "are not so different from prior ones . . . in the end, it really is still all about profitmaking and achieving overall capitalist economic goals . . . They colonize spare time and transform free time into labour time, in which surplus value is created and appropriated by capital. However, the prosumers don't realize that they are being exploited because exploitation now seems fun to them; it is entertaining, and takes place during their spare time." These type of arrangements (such as those of the digital economy, or co-creative relationships between game player and game creator) may also be more problematic because they could lead to more precarious employment and less security, more unpaid labor and more exploitation, as well as the use of contract workers and freelance workers who are more easily expendable. Fuchs explains that, "in reality, Web 2.0 both affirms capitalism and produces potentials that can undercut profitability and anticipate a fully cooperative economy. In this respect, Web 2.0 is characterized by the antagonism between the digitally networked productive forces and the generalized capitalist relations of production."[52]

What aspects of crowdsourcing, citizen science, or knowledge gaming might be exploitative, and can we compare these alongside other traditional forms of exploitation, such as the sweatshop or factory model? Often, people engaging in crowdsourcing or game content creation may not feel like their labor is being exploited. In fact, they do not feel like it is even work. Banks and Humphreys researched when the language of exploitation begins to be used by game players and game developers.[53] The typical assumption is that game players are easily and unknowingly taken advantage of by the game developer and corporate interests. However, the results of their study suggested that this is not the case, and game players were very aware of their relationship with the game developer, the labor of their content creation, and the requirements of the game content's production.[54] If game players are aware of these relationships and still participate in them, are they problematic? Andrejevic asks whether we should be concerned about the exploitation of players that does not seem to lead directly to immiseration or "the aggravation of human misery."[55]

Nick Couldry argues that while activities, practices, and relationships may not *feel* exploitative, "they create asymmetries that are preconditions for future exploitation or abuse."[56] The design or use of a knowledge game may not be inherently exploitative or cause misery, but it could create conditions where exploitation can occur, even far into the future. The system or structure that is designed may incentivize coercion, for example, which could lead to problematic exploitation eventually. Coercion could be entrenched in the design of the game system or even unwittingly part of the sociocultural context that emerges around the game. Can we conceive of design principles that might support (or subvert) coercion or exploitation?[57]

At the same time, volunteering and many types of activist work have always been unpaid, yet this labor improves humanity and contributes to the common good. Would we want to stop people from volunteering because it could potentially be exploitative or coercive in the future?

Is there a hybrid category between free or voluntary labor and paid labor? Campbell recalls Kuehn and Corrigan's idea of "hope labor" or Marwick and Duffy's "aspirational labor," suggesting that some people may be participating or contributing labor so that they can access paid jobs or tasks that help them achieve compensation.[58] While it is unclear whether people are participating in knowledge gaming because they really want to be scientists, social scientists, or researchers, or if they are just hobbyists or game players who want to participate, would we want to stop people from building the skills and knowledge to achieve professional and personal goals?

This brings up more questions. If the work done in a knowledge game is exploitative but benefits society greatly—such as by helping to cure cancer or stop global warming—is that exploitation acceptable? Is it less exploitative if the same type of knowledge game, with the same type of exploitation of labor, contributes so-called good knowledge versus knowledge that can be used for reprehensible purposes? Is it acceptable if the game contributes knowledge that increases profits for someone else, such as if the HIV drug created based on the *Foldit* contributors makes millions of dollars for a drug company or if a cancer research firm develops profitable interventions that help cancer patients based on data analyzed through *Play to Cure*? What if these profits later go toward organizations supporting genocide or terrorist activities?

Campbell, citing Terranova, makes the distinction between volunteer work and digital labor that is free by arguing that "volunteers and activists are aware of who (at least abstractly) benefits from their efforts." The asymmetry and imbalance of power occurs when the laborer gives away her work for free but does not know who benefits from it.[59] Thus, alienation may be a good term to adopt and apply, but how does alienation function in knowledge games?[60] When are people alienated from their original purpose in playing a game, such that they serve another's purposes rather than their own or the one they expected? If some people (say, scientists) secure more gainful employment because of research published using the digital labor (e.g., data-analysis tasks) performed by the public, is this an unfair advantage that exploits and alienates the knowledge game contributor? Published research, though beneficial to the public, may also be extremely valuable to the originator of the knowledge game. What if the data analysis generated by the public gets used to generate a new, profitable invention? Those profits could, in the future, be used to support causes that the public had not intended to sustain—indirectly alienating their work from its original intent. How do we ensure that this type of "work" stays in the volunteer sphere and not in the potentially exploitative or alienating sphere?

Moreover, if we label the contribution done in a knowledge game as work or labor, at what point do we need to ensure the appropriate treatment of its players, and when does it become further exploitation if we, as a society, do not consider how to care for and protect them?

Toward a New Complexity

So is it all doom and gloom or happy-go-lucky optimism for knowledge gaming? What are the implications of these critiques of digital labor and participation for knowledge games? Should we protect people from participating in these games? Or should we encourage them because they lead to innovation and more broad participation in knowledge production?

On the one hand, Walter Benjamin argues, "The more consumers become producers, the better a media apparatus like the newspaper or the theatre would be." I described in chapters 3 and 5 the benefits of including more voices, more perspectives, and more problem-solving strategies to gain optimal solutions, peak idea generation, and more innovative knowledge production. In this chapter, I also described participation models based on mutual benefit among producers, consumers, and owners, and what Jenkins calls a type of "affective" or "moral" economy where participants can contribute meaningfully in participatory cultures, or share knowledge in affinity spaces.[61]

New connected technologies, such as online knowledge games, could more easily support an exchange of ideas, access to means of production, or the co-creation of knowledge. On the other hand, I also described how the new digital economy has brought to the surface potentially problematic relationships between audience and owner, consumer and producer, and a new type of immaterial or indirect labor that is often free and undervalued but is used by capital for its surplus value. This leads to more questions. Is there merit to both perspectives? Is there really more freedom with digital technology? Or are we more trapped?[62] Are we more democratically engaged through these new forms of participation or are we more subjugated?

The interconnections among digital labor as participation, empowerment, expression, and liberation versus exploitation, immiseration, alienation, and coercion are complex. In this chapter, we have seen co-creation as either an "extension of market exchange, and thus a product of incentives associated with existing institutions; or view[ed] consumer co-creation as the emergence of a new non-market model of production centered on sociocultural explanations."[63] While it may appear that there are two camps— those that are more optimistic and those that are more critical of digital participation—the reality is more complex.

Postigo argues that one way to evaluate co-creation is to look at the design and architecture of the technology—in this case, the knowledge game itself—as well as how its design is refined by the play of the game and its players and the social systems and practices around the game, as sometimes unseen design aspects and "'natural' elements of the participatory/labor space actually serve as strong influences of action." Postigo analyzes the social and technical affordances generated by YouTube and how these affect the use of the platform and how it fits into culture. Affordances are types of functions and activities enabled or allowed by a particular design, platform, or technology.[64] A technological affordance of knowledge gaming (and *Happy Moths* specifically) may be to enable more people to participate directly in the creation of the scientific knowledge of moths by being able to categorize moth images from the comfort of their own homes. A social affordance of *Happy Moths* could be the reshaping of the relationship between entomologists and amateur bug collectors.

Likewise, Banks and Potts look specifically at modding in *Fury*, a player-versus-player MMOG, to understand how co-creativity emerges. They find that co-creation is voluntary in *Fury*, meaning it is not coerced or incentivized in some way but has subtle forms of incentives, such as "reputation, opportunity, learning, recommendation, and access." User participation or co-creation is not just about providing cheap or unpaid and exploited labor. Rather, Banks and Potts observe that co-creation forms a dynamic relationship between commercial and noncommercial, individual and corporate, interests, rather than one side winning singlehandedly.[65] The researchers are not merely optimistic or pessimistic about the role of co-creation; rather, they acknowledge its messiness.

Perhaps we need a new model that incorporates all these perspectives to disentangle the messiness of coercion, motivation, and agency in co-creation. This new model would explore how forces affect each other—both the "market-based extrinsically motivated exchange relations and culturally-shaped intrinsically motivated production relations," which Banks and Potts consider a co-evolutionary mix of both economic and cultural factors. Understanding this dialectic will become more and more pertinent and should not be underestimated. As Banks and Humphreys explain, "ordinary consumer creativity is rapidly becoming an increasingly integral part of the global market capitalism model of production"; likewise, ordinary game player participation in knowledge games may become more essential to global knowledge production.[66]

In the next chapter, I explore another dimension of knowledge production—using data and data-analysis tools to create new knowledge. What can we learn from data, and what are the limits?

CHAPTER EIGHT

Data

In *Play to Cure: Genes in Space*, players whiz through space, plotting the best route for their spacecraft to get as much Element Alpha as possible. Element Alpha does not really exist, but it is a highly sought-after substance in this fictional universe. Element Alpha does have a real-life counterpart, however. It represents actual genetic micro-array data from breast cancer patients managed by Cancer Research UK. Cancer Research UK has a huge backlog of breast cancer data and not enough people to analyze it—it's just too much for one individual or small team to sift through. A computer cannot process the data because unlike a computer, a human being can quickly identify where the dots of cancer data are denser. By looking at Cancer Research UK's data set, a player can identify the denser areas that differ from the rest of the data pattern—the anomalies—which helps scientists understand more about breast cancer.[1]

This is where *Play to Cure* comes in. According to the lead designer of the game, the player's "spacecraft is flying 'over' a graph that has been rotated 90 degrees, and we are recording the position of the spacecraft in order to analyze the data."[2] The route a player uses to drive her spacecraft in the game shows the scientists where concomitant data densities and anomalies lie in the data. In other words, where the person flies to get more Element Alpha can clue in scientists to the density anomalies in the cancer data. Thus, people are playing a game about space adventure, but they are really analyzing breast cancer data and, by doing so, contributing to our collective knowledge about breast cancer.

What does it mean to produce knowledge through the creation, use, and manipulation of data? Many of the current knowledge games and citizen science projects have centered on generating data, analyzing data, and on solving problems and answering questions through the processing of large amounts of data. In this chapter, I explore the potentials and limits of data, including data collection and analysis, as applied to knowledge games.

What Is Data?

There is a mythology around data, or the "units or morsels of information," that the more of it we can access and analyze—about ourselves and about each other—the more we will be prepared, ahead, and evolved. In an oft-cited 2008 *Wired* article, Chris Anderson portended "the end of theory" and the obviation of hypotheses and the scientific method, arguing that we will be able to prove everything and anything if we just have enough data. Models and theories are no longer necessary because they are replaced by a massive amount of information that provide proof without a whiff of doubt.[3]

Thus, the era of "Big Data" had been heralded. Since then, people have espoused data's ability to help cure cancer, solve hunger, understand depression, end war, and keep more kids in school. The notion is that if enough data are collected, particularly now that we have access to a wide array of data collectors, such as wireless sensors, personal mobile devices, FitBits, smart refrigerators, and drones, then perhaps we can learn more about cities, physical fitness, access to healthcare, food safety, prisons, terrorists, and our entire world. Maybe we can predict the future.

Even the White House is getting big on data. Under the Obama administration, Big Data is now a $200 million initiative, which aims to "extract knowledge and insights from large and complex collections of digital data" and promises to enhance our understanding of science, engineering, teaching, and learning, while also empowering national security. But what is Big Data and should it really be so "big"? Is Big Data—and how we use it—transforming society and contributing to an "information revolution"?[4]

The term "Big Data" initially referred to the amount of data that could not fit in computer memory or, more generally, to an amount of data that is just beyond the storage and analysis capabilities of an organization. Edd Dumbill defines Big Data as data that surpasses the ability of a system to process it.[5] Based on this definition, then, Big Data could refer to almost any amount of data, and the actual quantity is more about what the purveyor can handle. So, this means that just beyond the amount of plant systems data that Mattioli and the amateur botanists could manage in their commentaries might have been considered Big Data in the 1500s. And just beyond what Cancer Research UK can barely manage could be considered Big Data today. The limits of data have expanded radically, but the concept of data being too much to handle is nothing new.

The term Big Data has recently become hot, in part, because the amount of data that is generated and collected has skyrocketed enormously. There's so much more data being created and recorded today than ever before. Zwitter and Hadfield explain that "2.5 quintillion bytes of data are created every day through pictures, messages, gps-data, etc." The amount of data collected from antiquity to 2005 is only one-tenth the amount created between 2006 and 2011. That total will continue to multiply, quadrupling an estimated every three years.[6]

It's Not Just the Size That Matters

But is this conception of "big data" really so different from what has been going on for thousands of years, only with a massive difference of degree? Is there simply a quantitative shift, in that we can now look at and process much more data than in the 1500s, or is there also a qualitative shift in how we use this data?

Lyon argues that there is a degree change in data but *not* a kind change. Andrejevic, however, disagrees. He maintains that there is a qualitative change from "the 'emergent' character of new data-mining processes, which can now generate un-anticipatable and un-intuitive predictive patterns . . . That is, their systemic, structural opacity creates a divide between the kinds of useful 'knowledge' available to those with and without access to the database." Thus, there are new techniques that unearth patterns that could not have been seen before. In addition, the perception of data itself has shifted throughout history. Puschmann and Burgess, referencing Rosenberg, write that the "sense of data shifted during the 18th century from anything widely accepted as given, granted, or generally known to the result of experimentation, discovery, or collection."[7] People, when conceiving of data, have focused more on its quantitative aspects but also on its need to be scientifically vetted.

Furthermore, data scientists often point to the fact that Big Data is not just about the large amount of information that exists but how it is processed, the processing speed used, and "the heterogeneity of data that can be dumped into combined databases . . . volume, velocity and variety," often called the "three Vs" of data analytics.[8]

Recently, new technologies (such as Hadoop's storage capabilities) (volume), as well as new data methodologies and analysis tools have helped data scientists find novel ways to handle heterogeneous data (variety), process it, and quickly find patterns (velocity). For example, according to Chen, Mao, and Liu, in the past, classic analytical tools included regression analysis (a tool for discovering relationships between an independent variable and possible dependent variables), correlation analysis (a method for finding relationships between phenomena), and cluster analysis (which involves grouping objects based on shared features and putting them into categories). Now, new tools and methodologies have been applied to data, such as hashing (transforming data into smaller, more quickly read, values), Triel (applying common prefixes to reduce the time of comparison on character strings), and parallel computing (independently co-completing parts of a task), all of which support efforts to manage the three Vs of data.[9]

Thus, there have been analytic and technical changes in how we process data but also paradigmatic changes in how we are using data to create and accept new knowledge.[10] In the social sciences, we often produce and establish knowledge through the collection and analysis of data. We may test hypotheses against theoretical frameworks or use representative sampling, surveys, or interviews. Yet with Big Data, there seems to be less of a reliance on the vetted social science methods.[11] Cukier and Mayer-Schönberger describe a move away from causality in lieu of strong correlation, suggesting that we only need to uncover correlations (that something is happening) rather than consider causation (why something is happening) to be able to make claims about the world and the relationships among variables. In other words, we can look at the grade on someone's first test in college and judge the likelihood that the person will remain a student at that college, based on the results of analyzing a large data set. We can look at a seemingly innocuous variable such as what type of outfit someone wears to an interview and potentially make data-driven, correlational judgments on whether the person would be a good fit for the organization.[12]

Moreover, Cukier and Mayer-Schönberger argue that we no longer need to tediously test hypotheses or provide caveats about correlation; with more data, a testable hypothesis is not necessary anymore to make conclusions. As Anderson claims, "With Petabytes [of data], correlation is enough."[13] The assumption is that the more data we get, the merrier, and the more variety being added to that data, the better as well. This process is far removed from the oft-repeated phrase in research that "correlation is not causation" or the idea that simply knowing two variables are related does not mean that they are causing each other. A third variable could be affecting any changes.

One reason *Play to Cure* is potentially effective is that it invites so many people to analyze breast cancer data. In addition, it seems that the game's designers assume that, whether having more perspectives on the data or finding more optimal routes through space to get the most Element Alpha, more is better. Likewise, the impetus behind InnoCentive or *Foldit* is that there is wisdom in the crowd and intelligence revealed in the collective opinions, expertise, and abilities of a large number of people, tools, and systems.[14]

The creation of trustworthy knowledge perhaps relies less on one person and more on an efficient data-processing system of people, places, and things. New knowledge production is about "not just giant computers but a network to connect them, to feed them, and to make their work accessible. It exists at the network level, not in the heads of individual human beings." This is similar to Gee's notion, described in chapter 5, that knowledge can be stored and distributed throughout a system of tools, objects, people, platforms, rather than just one person's head. These assumptions underlie the promise of Big Data, with a datum standing in for a unique perspective or brain, and data mining and analysis pledging to help us access and interconnect as many of these minds as possible and create the best new knowledge we can.[15]

So What Can Big Data Really Do?

Big Data popularly refers to not just the amount of data available or capable of being processed but also what we can do with those data—"things one can do at a large scale that cannot be done at a smaller one, to extract new insights or create new forms of value, in ways that change markets, organizations, the relationship between citizens and government, and more."[16] With these supersized expectations, it is no wonder that Big Data is perceived as so powerful and a panacea for everything from environmental catastrophes to genocide.

The assumptions about the power of data do not just underlie the pursuit of knowledge in the sciences, as in Cancer Research UK's *Play to Cure* project, but they are also the driving forces for newly hot fields in social sciences and the humanities that merge data science with not traditionally data-driven or scientific fields. Cukier and Mayer-Schönberger observe that Big Data is infiltrating all areas of humanity. Data, Puschmann and Burgess argue, "can be effectively harnessed to better approach a wide range of societal issues, from economic growth and development to security and health care, with far-reaching implications." For example, the field of computational social science focuses on using large-scale data-driven methodologies to understand how people behave individually and in groups. We also now have data-focused history and literature research initiatives, and historians

are being urged to study programming so they can better use tools to algo-
rithmically address, for example, thousands of historical documents, rather
than just the amount they could read through on their own.[17]

In *Social Physics*, MIT Media Lab researcher Sandy Pentland explains how
Big Data can help us not only better understand and predict human behavior,
but even shape it. And Ben Waber, in *People Analytics*, shows how real-time
data from sources such as e-mail coupled with environmental sensor data
helped Cubist Pharmaceuticals change the layout of their office to increase
social interactions among employees, and ultimately, enhance profits.[18]

Right now, large-scale social science data crunching is typically the domain of large companies, such as Google or Zynga. Facebook examined data
from users' newsfeeds to understand (and manipulate) how the emotion-
evoking words used in friends' posts might translate to users' own emotional
expressions in their own postings. In other words, Facebook researchers
found that when you read sad stories in your feed, you are more likely to then
use sadness-evoking words in your own postings, which may even affect or
be indicative of your own emotional state. Likewise, the Game Show Network analyzes social data from its games to optimize its game incentive
structure and customer acquisition strategy to generate profit, as do many
social online "free-to-play" games.[19] More benevolently perhaps, game companies may use player data to alter maps and provide extra wayfinding and
signage to help players move around a game world more efficiently, though
enhancing profits from user satisfaction is the ultimate goal in using and
analyzing data.

Should top-notch data methods only be in the purview of wealthier prof-
itmakers and used to meet a corporate or private research agenda, or should
state-of-the-art data collection, analysis, and mining capabilities be opened
up to be used to enhance knowledge for the common good? And why not
apply these methodologies to games? Why not use knowledge games to collect and analyze social data? We could use them to support computational
humanities and social science research projects and make it easier to generate,
collect, annotate, and analyze the past, the present, and our human condition.
Since organizations are already collecting from so many data sources to
better understand human (or consumer) behavior, such as from sensors,
Facebook posts, tweets, and Web content—what about also creating games
with the purpose of investigating game-playing data and in-game behaviors to generate new, socially beneficial knowledge about human emotion,
behavior, and attitudes? For example, if we gather enough data through a

game about the activities, actions, and beliefs of depressed people, we could conceivably use that information to make predictions about how depressed people might act in a future game and then use the game to intervene or to adjust their behavior.[20]

Or consider a research project completed by the Open Academic Analytics Initiative at Marist College, which used data analysis techniques to detect predictors that could be used to capture those college students who were at risk of dropping out or not succeeding in college.[21] Could we collect these kind of indicators through a knowledge game and even reshape the game to address those risks, such as through the use of in-game peer-to-peer tutoring on targeted problematic academic or skill areas?

Limits and Potentials of Data

Based on the aforementioned examples, it seems like a no-brainer to use a variety of platforms, such as Facebook, Twitter, various crowdsourcing or citizen science projects, and even knowledge games to generate and analyze data and use that information to better our understanding of the world. But is this really an effective way to produce knowledge? What is gained or lost when we start to base knowledge on data, whether in a game or not? How might complexity be diluted when we use data?

Just as we saw with participation, the potentials and limits of data are complex and often contradictory. Will Big Data help us cure cancer, combat depression, figure out global warming, and address world peace, or will it will seep into our unconscious, provide the ultimate panopticon into our lives, decreasing our freedoms and increasing institutional control over our lives? In "Critical Questions for Big Data," boyd and Crawford ponder, "Will large-scale search data help us create better tools, services, and public goods? Or will it usher in a new wave of privacy incursions and invasive marketing? Will data analytics help us understand online communities and political movements? Or will analytics be used to track protesters and suppress speech? Will large quantities of data transform how we study human communication and culture, or narrow the palette of research options and alter what 'research' means"?[22]

What are some of the limits or challenges with not only collecting and storing but also with using big data sets? First, there are technical considerations, such as processing speed or storage capacity, which are described in detail in a number of articles and beyond the scope of this book. But are there issues with data even if no physical or technical barriers exist? Boyd and Crawford offer six statements or "provocations" that should be considered when using Big Data.

The objectivity and accuracy of Big Data may not be what they seem. There is a myth about Big Data, argue boyd and Crawford. Qualitative researchers are viewed as merely interpreting and telling stories while quantitative researchers deal in hard facts and figures.[23] As described in chapter 6, science is seen as objective, legitimate, and based on solid data, whereas social science and humanistic perspectives are seen as less rigorous and softer.

However, in actuality, all data is subjective and, as such, subject to biases and assumptions, even scientific data. Researchers are arbiters and interpreters of data, which means that each discipline has its own ways of framing data and using it to create models of the world. Numbers are not neutral. Which questions get asked, which variables are reviewed, how the data get shaped, cleaned, and filtered also affect how the data are interpreted and used. Data analysis is always limited and biased, soft and pliable. Data always come from somewhere, and the questions, decisions, and assumptions that lead to their collection and usage also affect data.[24]

Bigger does not mean it is more accurate or higher quality. Just because there are lots of data does not mean they are accurate, representative, or appropriate for measuring what we want them to measure. Lots of bad data are still bad data. Moreover, lots of data can show patterns where none really exist, such as correlations that are erroneous or nonsensical.[25] Likewise, let's pretend we want to use a game to conduct a study of a sample of the population. Even if we create a game that houses thousands of players, that sample may not be representative of a population as a whole, or the population we want to study. And when we couple this with the fact that of the players, only 1% of them may be supercontributors who are providing most of the research data, it could skew the results even more.

Further, biases that affect data can be introduced at any stage of research.[26] No research is bias free; we just have to be aware of biases and evaluate conclusions in light of them. In social science research, potential biases that can affect data collection and analysis include selection or sampling bias (where the people in the study may not actually represent the population from which it was drawn) or information bias (where the way the data are collected is misleading or erroneous), among others.

In the case of InnoCentive, we learned that the platform's population is overwhelmingly male, older, and well educated. Therefore, we cannot necessarily trust efforts to use studies of InnoCentive participation to make claims about how people solve problems collaboratively in general. Likewise, in a game, there could be biases introduced based on who actually plays a given game (self-selection bias), how many games people have played before (lead-time bias), only collecting data on players who complete the whole game (and not those who start it and then immediately leave), what types of data are collected for a game player, or the ways that any data are categorized.

Already we can see how easy it would be to introduce biases into even a clear-cut game or crowdsourcing exercise. Having more participants or more data from those participants does not necessarily remove the biases completely. While some statistical methods can diminish the effect of various biases, they may not be able to erase it entirely. Thus, increasing the size of a data set—having more erroneously collected and biased data—does not reduce the amount of bias that it has.[27]

Besides introducing biases, we can also make inappropriate or inaccurate interpretations of data, and draw incorrect conclusions. In some cases, there can be a confounding variable (such as a third, untested, or not considered variable) that affects the behavior or outcomes of people in a study. For instance, data may suggest that there is a relationship between eating food and gaining weight. A likely conclusion we could draw, then, is that if we eat food, we will gain weight, and that there is a causal relationship between the two variables. But is this a causal relationship or a correlational relationship? Is this is the only direction—that food affects weight gain—that we should consider? We can also consider the opposite direction—that if we gain weight, we might feel depressed about it and then want to eat more food. Or there could be a confounding factor. Let's say the person gaining the weight has an underlying thyroid issue or genetic predisposition to gaining weight. Maybe the person has a sedentary lifestyle that is factoring into the weight gain. Or maybe the person is pregnant with twins. This is a simplistic example, but it illustrates why it is difficult to draw clear conclusions from correlations, even with tons of data.

Big Data should not be decontextualized. We cannot just make claims and models using data without thinking about its context, and how that context provides nuance and meaning to the data. As Gitelman and Jackson contend, no data is raw, or free of any social context.[28]

To illustrate this, we can look at the language used to describe Big Data. Is Big Data (or any ol' data) really so "powerful" and "all-knowing?" Can we "unleash" it like a lion? Is it so big that we need to "tame" it, like a "beast"?[29] The very way we talk about data colors how we use it and what we can do with it. We cannot assume that data is free of social negotiations and expectations.

If we lose the context of the data or misapply it to new contexts, we can also lose the meaning of that data, which could lead to inaccurate conclusions. For example, if we investigate problem-solving strategies in a game and find that most players get stuck solving a puzzle that uses the word "jejune" but not the puzzle with the word "waterfall," we might conclude that the players in the game are deficient at understanding or spelling the word jejune. However, if we look more deeply at the context of the game, we might find out that the jejune puzzle was missing a letter due to an in-game bug or it had a particularly difficult "boss" protecting it, rendering the puzzle harder to solve—rather than the word itself being the problem.

Decontextualizing data also brings up other issues, to which the consequences are still unknown. On the one hand, *Play to Cure* may lead to important discoveries about breast cancer based on the cancer data players analyze in the game. On the other hand, the game obfuscates the origins of the data by transmogrifying it from actual people's breast cancer data into the imaginary Element Alpha. Those playing the game are separated from the original source or context of the data so that they are progressing in a game rather than thinking about analyzing cancer data. Instead of considering the people who suffered from cancer, they are merrily racing through space. The game not only trivializes and conceals how the data were drawn but completely reconceptualizes their use. This has implications beyond even a lack of understanding of where one's data came from because it also represents a lack of understanding of how the data is being played, and of what that means for the play experience. What are the consequences and ethical ramifications of this type of distortion of context?

Ethics matter. While data can open up new areas of inquiry, it also brings up many more considerations, particularly about ethics. Questions about ethics have underscored all of the provocations in this section, but they are also important to bring up on their own.

We may think that providing or playing data, particularly anonymized data, is free of risks, but this is not entirely true. Boyd and Crawford argue that anonymized data still has privacy and other concerns, particularly when it comes to relationships, social interactions, health, and economic considerations—or pretty much any data that is on, about, or surrounding human beings.[30]

We hear a lot about privacy issues with our personal data, especially in relation to companies such as Facebook who randomly adjust privacy notifications and use personal data for their own corporate purposes. But the concept of privacy is complex because, as we see, many people skip reading and reacting to the privacy notifications and agreements on sites (or even doctor's offices or company workplaces) and yet still care about how their data is used.[31]

For one, people *do* want to know how websites and other platforms use their personal data, even if it does not seem like it, because they skip over privacy notifications or divulge that data so easily. Andrejevic notes that in a study of over 1,100 Australians, 73% of respondents did want to understand more about how websites collect and use their data.[32] As Kosciejew argues, data may start by being used for one purpose and then be sold and reused for something entirely different. People might agree to one use of certain data and not be aware of all the other potential uses or future applications, such as their cancer data being used in a spaceship game, or their mental health data being used in their workplace. Thus, people should only be encouraged to supply data after fully understanding how their data could be used, or how inheritors of that information might rely on it to make decisions about them.[33]

In fact, it's not that online and game participants do not care about privacy—rather, it may be that they feel powerless about their options. Andrejevic conducted interviews related to Australians' attitudes and reactions to the collection and use of personal data and found that the people he interviewed felt cajoled to share information and did not know how it would be used, which in some ways felt worse than the idea of other people simply knowing their personal information. The interviewees expressed their feeling of disempowerment and lack of understanding about how the data would be used and the consequences of their data submission, almost as if they were oppressed by those using their data. The powerlessness they described, it seems, came from both a lack of control over the resources and platforms they were using to provide their data, as well as their inability to make decisions about or process the knowledge generated from the data.[34]

So what about the cancer data set that is being used in *Play to Cure*? Did patients know that their data would be transformed into a space game? Would they have had the resources or abilities to analyze, compare it, and draw conclusions from it anyway?

The institutional research board (IRB) at a research institution such as a university is typically the stopgap that helps researchers secure and safeguard data and use the data appropriately in any type of human subjects research. But what happens in knowledge games or even corporate-led platforms? Is Facebook using an IRB to make public claims about its newsfeed manipulations or for internal ideas about how to better target certain demographics? Are people's contributions in a knowledge game part of the research process? Or is it just game playing? What type of privacy is surrendered (and not protected) when we play knowledge games? Should knowledge game be required to follow certain safeguards?

What happens if data from a knowledge game are used in a way that a player does not expect? For example, let's reconsider an imaginary game given to all incoming college first-years that touts itself as helping users learn more about college readiness but is also silently used by the designer to identify potential college dropouts. What if a person playing this game does not realize that her game playing will flag her as having several "dropout" risk factors, which then require her to participate in time-consuming tutoring that will take her away from the job that she needs to feed her new baby or her sick mother?

How do privacy and informed consent policies matter when applied to games?[35] Is it the researcher's responsibility to protect the player—or even the game designers, developers, promoters, or marketers, and to see that any policies are followed through—and to encode such policies in the design of the game, social media marketing, or backend storage and access systems?

Although the IRB process is not flawless, it exists to question practices around data usage that may not get questioned otherwise. "Researchers must keep asking themselves—and their colleagues—about the ethics of their data collection, analysis, and publication," argue boyd and Crawford. People, whether knowledge game designers, players, or researchers, need to think about the broader implications of data collection and analysis. What are the full consequences, and how are players and society more broadly protected?[36]

The data divide matters. Boyd and Crawford and Andrejevic have described a difference in power between the data-haves and the data-have-nots. There are those who can access and use Big Data, with its required storage facilities, algorithms, analytical tools, and computational abilities, and those who cannot. Facebook and Google have access to the data they collect from their users and decision-making power over what to do with it, while users do not.[37] There is privileged access to data and how it is used.

More specifically, Andrejevic argues that just as there is a "digital divide" between people who have computers and access to the Internet and those who do not, there is also a data divide. This divide goes beyond some researchers having access to data and some who do not. Rather, there are also "those who are able to extract and use un-anticipatable and inexplicable . . . findings and those who find their lives affected by the resulting decisions." Even if people have access to data, they may not have the ability to crunch meaningful patterns, compare it to other data sets, or use it in a way to predict their own futures; nor do they necessarily have the ability to store this data. This divide creates what Manovich calls the new data classes: those who contribute the data, those who collect it, and those who can analyze it.[38]

The concept of the data divide speaks further to the power differential between researchers and participants in citizen science projects or knowledge games. Are researchers similar to corporate data miners, asking disempowered people to contribute their data for the more privileged data classes to utilize and profit from? Not only do the researchers have the expertise to know what to do with the data, but they also have the tools and capabilities to do something with it, through their access to "databases, storage, and processing power." But it's not just about controlling the databases or owning the game; these researchers also have the social capital, trustworthiness, and value to turn what they discover from the data into accepted and established knowledge.[39]

What are the implications of an unequal division of power stemming from this Big Data divide? First, the exclusivity of the data domain overvalues the contributions of data to knowledge and those who can analyze it. It sets up flawed dichotomies between, for example computer scientists and social scientists, with the former knowing everything and the latter perceived as not having anything to provide, further contributing to the current mythos of the computer or data scientist (and concomitant downplaying of the humanities). We see these misperceptions played out more broadly and societally in the types of jobs that are offered, who is valued and/or compensated better, and the types of college majors parents want their kids to have.[40] Likewise, in K–12 schools, people are pushing for STEM (science, technology, engineering, and mathematics) and computational-thinking skills, whereas there is currently less pressure for the inclusion of social science analysis, theory, and methodology, or humanistic inquiry. These preferences affect what questions get asked and what types of knowledge gets created. If computer scientists' and data analysts' perspectives are deemed the most important, they end up being the perspectives that are reflected in new knowledge.

Second, the data divide also ties in with issues of surveillance, in that the concept problematizes who controls data and whose data gets controlled. How might organizations and institutions be using data as an instrument of control? Andrejevic argues that those with access to processing power can use that power for disreputable purposes, such as social control or for reinforcing social differences—"surveillance as social sorting." Gandy argues that this type of social sorting is not necessarily just about sorting by people's preferences or backgrounds, such as who they are or what they want, but it could be used to predict what they will do, which is a form of surveillance.[41]

Finally, Andrejevic argues that the data divide gains competitive advantage for only a minimal few. "The asymmetry and opacity of a 'big data divide' augurs an era of powerful but undetectable and unanticipatable forms of data mining . . . [It] privileges a form of knowledge available only to those with access to costly resources and technologies over the types of knowledge and information access that underwrite the 'empowering' and democratizing promise of the Internet." In other words—and here's the real kicker—the so-called democratic and empowering Internet enables people to more readily participate and share their data and be part of the knowledge conversation. But that sharing and contribution ironically promotes more unequal access to knowledge because it contributes to people being further controlled.[42]

If we all do not offer our data and perspectives to help contribute to finding a solution to a human problem, does that constitute less of a democratic engagement with a given problem because the perspectives do not come from a representative cross-section of the population? In other words, do we need more people to participate to make problem solving more relevant, representative, appropriate, equitable and democratic? If so, do we then simultaneously reduce our privacy and personal control and, therefore, the equality of our participation? These are questions that are not easily approached.

The nature of knowledge changes. Does Big Data also change the nature of knowledge and how we arrive at knowledge? Boyd and Crawford argue that Big Data changes what it means to research or what it means to know. It changes how we collect and analyze data, not as it relates just to scale but to epistemology. "Big Data reframes key questions about the constitution of knowledge, the processes of research, how we should engage with information, and the nature and the categorization of reality."[43]

Boyd and Crawford use as an example the design and structure of Twitter and Facebook and how it is difficult to archive or search data in these platforms. As a result, the questions that get asked and studied are typically more immediate reactions to short-term events or happenings. Right now, using Twitter for example, it's easiest to analyze event-based, short-term data from the Super Bowl or *Bachelor* finale, or a tsunami, blizzard, or big election results rather than to look at older or longer-term data, such as comparative data on incremental changes in viewpoints over the course of a political campaign.[44]

Likewise, the affordances of knowledge games will help decide which types of research questions get "well played" through games. The limits and boundaries of games will inherently affect and dictate the outcomes and types of knowledge that get produced from them. Right now, it is easier to have game players categorize images or provide quick answers to simple questions (such as *Who Is the Most Famous?*) than to interact more deeply with others in a complex social situation. It is easier to create cooperative contribution games rather than algorithm construction or adaptive-predictive games (see appendix A), so the former may be more likely to be made. Certain game mechanics or platforms are easier to design with and for, so these end up helping to decide the types of information, data, or actions people can play in knowledge games. Thus, knowledge games will not just measure and observe objectively; they will also shape our reality, our social relationships, our ways of solving problems, and our ways of knowing.[45]

A Little Data

What, then, is the value of data? Do we really still need theories, models, and the scientific method, or should there be a paradigmatic change in how we create knowledge?

As we struggle with the complexities of collecting and using data, and the types of knowledge we can create with it, we must also grapple with questions about what knowledge games, particularly ones that collect and use data, can and should do with these data. If we are using *Play to Cure* to help cure cancer, can it really be so bad if data is transformed or separated from its original intent? If the data generated and knowledge produced end up curing breast cancer, does it matter how the data were played, whether they were bias-free enough, or whether the scientific method was applied appropriately?

Andrejevic argues that although patterns might emerge in large data sets, we still need to be arbiters and interpreters of data—we need experts who can make meaning out of the patterns or even decide which patterns to look for. Responding to claims that correlation is enough, he writes, "Perhaps a particular combination of eating habits, weather patterns, and geographic location correlates with a tendency to perform poorly in a particular job or susceptibility to a chronic illness that threatens employability. There may not be any underlying explanation beyond the pattern itself." In other words, we may know that there is a relationship between two variables, but it may not be a causal or predictive relationship. We do not know if there is another variable affecting the outcome, or which variable is affecting the other.[46] We still need people who can interpret data and draw stories from it. We need people who can reshape data into actionable and meaningful knowledge and reflect on its implications and consequences.

Moreover, we need to consider what is lost or gained when Big Data analytics and data mining are our go-to methodologies. What are the hidden nooks and crannies of humanity that large-scale analytics and Big Data may miss? What can we learn from littler data? While computational and technological instruments can solve problems quickly or help us take action, they do not necessarily capture the nuances, the outliers, the extremes, the art, or the flavor of the human experience. Bowker imagines "living in a world where science and social science are about manipulating the world—effective action is after all a good thing. However, this is a massive reduction of what it means to 'know.'"[47]

In our haste to find ways to accommodate more data collection and analysis through games and interpret that data properly, let's not forget the standards, ethics, and perspectives gained through rigorous research methods, as well as the variety of methods available for moving toward understanding human truths. This does not mean that we should throw away Big Data; rather, we should be open to what it can and cannot tell us. We should not privilege certain types of methodologies for their contributory value, such as quantitative approaches in lieu of ethnographic investigations, interviews, close readings of literature, or one-on-one discourse.

Moreover, we should consider what is lost when we focus on a multitude of data rather than multiple paths toward knowledge. Croft argues, "Big data cannot account for the cumulative effect of a movement, it cannot perceive intensity of human emotion, and it cannot predict free will." It is the difference between sensing the mood in a room full of your family members and gauging how they feel only by looking at their posts on Facebook. Something gets lost and gained with each methodology, and there are benefits and limits to what we can know through each. Data does not replace human interactions or humanistic methods of knowledge creation.[48]

After all, books have been written from a single point of view and have spurred knowledge production and changes in society (some for better, some for worse). The case of Anna O., a single study of a woman suffering from hysteria under the care of Dr. Breuer, influenced Freud and helped inspire modern psychology. A singular game experience, such as *That Dragon, Cancer*, may be able to express a human truth about cancer that perhaps thousands of people analyzing breast cancer data in *Play to Cure* cannot. How can games make us care, doubt, empathize, express, and wonder in ways that may also support our understanding of the world?

What is lost when we consider data the only important text? A data set cannot express what literature, poetry, sculpture, discussion, or even a game can evoke.

Ultimately, data is not knowledge. In the next chapter, I continue to challenge and question the concept of knowledge and how this relates to knowledge gaming and democratic engagement.

CHAPTER NINE

Knowledge

Do you want to get the shrimp scampi or the spaghetti? What about the Caesar salad or minestrone soup to start? Are you serving the food? If so, make sure you get the orders right. Are you eating the food? Then don't forget to pay your bill at the end of the meal! It seems mundane enough. We've all gone to sit-down restaurants, waited for a host or hostess to seat us, looked at a menu and ordered food, interacted with the waitstaff, eaten the food, and then paid the bill. We know the drill because we have followed the "restaurant script" many times—a typical series of interactions that most of us learn, even from a young age.[1] *Sesame Street* episodes have explored Grover's poor waiter service at Charlie's Restaurant, which comically shows what happens when the restaurant script is not followed. This is funny, even to toddlers, but only because we have an established and shared understanding of the restaurant script.

The mechanisms for the creation and spreading of this restaurant knowledge among human beings seem to be effective. They include modeling and observation, such as parents modeling behavior to young children or adolescents observing each other; didactic expression of the restaurant's rules, such as an in-restaurant sign ("no shirt, no shoes, no service") or a message on a menu; and mediated illustrations of appropriate restaurant interactions on shows such as *Sesame Street*. Also, massive cultural, political, and economic forces underlie the institution that is "the restaurant" and the many related and overlapping systems, such as that of family, leisure, work, and business. The behavioral script involves both tacit and explicit expressions of norms and values, such as what it means to provide service and hospitality, to exchange money or transact, to share or not share, or what food represents, such as being a symbol of nourishment and love. And yet, all of these cultural, psychological, economic, and political conceptions of food and restaurants somehow get replicated and renegotiated and result in the deceptively simple restaurant interaction that many millions of people participate in every day.

But what if we suddenly go to restaurants with the new expectation that instead of providing money in return for the food, we need to step in and cook a meal or wash dishes? How would we learn this, and how would this

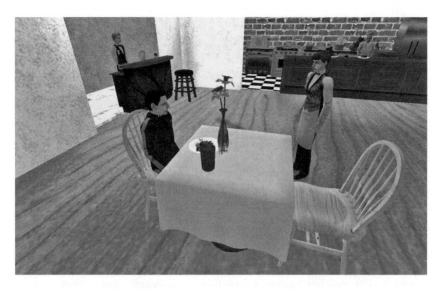

The Restaurant Game is a virtual environment where participants interact in a restaurant setting to "teach" the computer the typical interactions that people have in a restaurant. Source: Jeff Orkin, CTO / Co-founder GiantOtter

new idea spread? How would this alter the cultural, social, and economic substrate underneath the institution of "restaurant"? Understanding the mechanisms of change is important as we consider how to take the restaurant script, with all of its sociocultural ramifications, and share it not only with other human beings but teach it to computers as well.

This is the premise of *The Restaurant Game*, created by Jeff Orkin and Deb Roy of the MIT Media Lab in 2007. Its goal is to get enough players to interact in a virtual restaurant so that a computer could learn the nuances of human behavior in that setting and better generate artificially intelligent but more humanlike characters. According to the game's creators, "This experiment aims to generate AI behaviors that conform to the way players actually choose to interact with other characters and the environment; behaviors that are convincingly human because they capture the nuances of real human behavior and language."[2] This premise—that we can teach computers the subtleties of human behavior through crowdsourcing and gaming—is furthered in designer Jeff Orkin's more recent efforts. His company, GiantOtter, is creating *SchoolLife*, a multiplayer game where players teach the computer about social interactions surrounding bullying to more realistically present future bullying scenarios. The overall goal is to learn enough about the complexities of bullying so that the game will more authentically help players develop empathy.

What are the implications of using games such as *The Restaurant Game* and *SchoolLife* to produce knowledge not only for society but also for computers and other nonhuman entities?[3] What does it mean to know—and does this depend on whether it is a person, game, or machine doing the knowing? As we start to see more and more games designed for knowledge production, what are the implications? How does seeking knowledge through playing games change (or not change) the way knowledge is created, accepted, and spread?

In this chapter, I raise questions, uncover assumptions, and consider perspectives about how we know what we know, how knowledge becomes knowledge, and what the very act of making knowledge through games helps reveal about humanity and the world. We might agree that it is important in a game to create more authentic and more intelligent nonplaying characters (NPCs), or characters that are computer controlled rather than controlled by another human being, to help us develop more human skills (as in the case of *SchoolLife* helping us further develop empathy). But what are the implications of making computers more knowledgeable—or even more human—so that we can better learn from them and better produce more knowledge for ourselves?

Ways of Knowing

First, what does it mean to know something, and when does that knowing turn into knowledge? Historically, the creation of knowledge has been seen as the effort of specific individuals, even if this is not how it was actually generated. The myth of the singular "genius," such as Einstein, or a person who singlehandedly contributes to our knowledge, still persists.[4]

Karin Knorr-Cetina argues that we should not think of knowledge as embodied in an individual, discipline, or specialty only. Instead, it is determined by epistemic cultures, which shape and are shaped by historical, cultural, interpersonal, and political forces. She points to the concept of "knowledge societies" that "run on expert processes and expert systems . . . structured into all areas of social life." This seems similar in concept to the communities of practice, participatory cultures, or affinity spaces as described in chapter 7, or even the "Samba Schools" described by Seymour Papert, where learners work together to teach each other, share concepts, practice skills, and make collaborative discoveries. Knowledge does not reside in the individual; rather, it is found in the network of people, places, things, tools, machines, and computers.[5]

Knorr-Cetina investigates two case studies from science—experimental high-energy physics and molecular biology—to understand the "machinery" behind knowledge creation, and she analyzes how the two cases are different culturally. Rather than knowledge being embodied in a belief, application, or property (such as a website or game), she describes knowledge as a set of practices that occur "within structures, processes, and environments that make up specific epistemic settings."[6]

As I described in the last chapter, however, people often mistakenly equate data and facts with knowledge. They forget, ignore, or unintentionally neglect the social context or practices surrounding the generation, procurement, and interpretation of data. Even knowledge itself, once established, becomes divorced from its original context, the social and cultural practices that led to its adoption and the systemic and structural forces that resulted in its creation. Data becomes a commodity that is bartered and exchanged for its own establishment as truth. In other words, the value of data is not only in its content but also in its ability to serve as its own currency, which it uses to earn a spot as socially accepted knowledge.[7]

This way of conceiving data is nothing new. As Mary Poovey describes in *A History of the Modern Fact*, people tend to consider knowledge—particularly figures, numbers, and statistics—as unbiased, decontextualized, and impartial.[8] Chapter 8 reviews the different dimensions of data and data's pursuit as knowledge. For instance, what ends up happening is that data-driven paths to knowledge become privileged over other methodologies, such as case studies, ethnographies, interviews, or small-scale focus groups. Certain types of questions, lines of inquiry, and ways of knowing get valued and reproduced in lieu of others.

Each discipline may have distinct standards or values regarding what gets accepted as knowledge and how that knowledge is generated. What a humanist deems worthy to study and to know could be different from what a scientist cares to investigate. These differing voices enable a broader system of deliberation and dynamism, so that we can look at problems, possibilities, and prospects from multiple points of view and gain insights from their interconnections and disjunctures. As we saw in chapters 3 and 5, the intermingling of perspectives and collaboration among players may even lead to more effective solutions and spur superpowered insights. The very existence of a pastiche of perspectives may be useful for problem solving and knowledge production. For instance, relying solely on data-driven methods or using data to make claims about the world may ultimately result in myopic knowledge production. Privileging one line of inquiry or one way of getting to knowledge may be damaging to the value of the knowledge itself. When using and designing knowledge games, we need to be open to all types of inquiry and perspectives on knowledge making. We cannot let games and their limits, the trendiness of methods, or even our own personal inclinations toward knowing dictate how we design for knowledge production.

What Gets Accepted as Knowledge?

As I discussed in chapters 5, 6, and 7, knowledge is socially constructed and becomes accepted in part through a community or social entity. Community members may argue and arbitrate until they reach some type of consensus, at which point new knowledge is vetted and accepted. In the *Wikipedia* community, for instance, the editors and readers achieve consensus by verifying each other.[9] In *EteRNA*, players try out and submit different RNA design solutions, which then get compared and vetted through the system. The game's creators predetermine the value of these solutions by first stating up front the requirements for each mission (e.g., the design specifications) and then determining how well each player's design solution meets those requirements.

Complete consensus among participants, however, is not always necessary for knowledge production to occur. There are many instances where a majority of scientists might agree on a course of action or finding, but members of the general public do not agree. Consider the persisting debates over whether to vaccinate children, whether global warming is happening, or whether evolution exists. A 2013 article in the *New York Times* describes how medical and scientific facts no longer hold the same weight as they previously did and how, in fact, it is almost trendy to be critical of science.[10]

Thus, the context of knowledge, and whether it is politically expedient, socially preferable, and culturally acceptable, also relates to how well that knowledge will be established. These practices are nothing new and affect not just the knowledge "stories" we tell but when, how, and whether we want to tell them. In chapter 6, I described how Galileo's mistake of showing his work to Baldassare Capra affected how he shared subsequent findings with the public. Likewise, Darwin waited over 20 years to publish *On the Origin of Species* (1859), which housed his theory of evolution, because he was afraid that the public would accuse him of being an atheist and he did not want to be embroiled in the controversy. Today, his work is still debated and considered controversial.

The story surrounding knowledge production may be just as important as the knowledge itself. What happens, then, with knowledge games that purport to study scientific, humanistic, or social scientific questions? How do they need to change, adapt, or shift to be publicly acceptable and trustworthy, as I examined in chapter 6? How might these games affect how knowledge gets shared and established, or even withdrawn and hidden?

Who Gets to Know?

Who is allowed to create knowledge through games? The creation and use of knowledge is a power that is limited and restricted. Some people have the power to decide which research questions get asked or how funding on research is spent. Those privy to research results might also be able to dictate how knowledge is interpreted, communicated, or spread. Are new communication platforms and cultural conceptions about knowledge really helping to change who has access to knowledge making and knowledge spreading? Is knowledge production still closed to the public, or are we starting to open it up to everyday, amateur participation? As I described in chapter 7, more and more people have access to modes of production and distribution, such as creating and posting videos on YouTube or contributing to scientific knowledge by categorizing galaxies, bugs, or plants. Dickinson and Bonney argue that citizen science projects are a "microcosm" of broader changes in practice of science. Michael A. Nielsen contends that knowledge creation is being reinvented, with projects such as Galaxy Zoo, *Foldit*, and the open science movement (an organized effort to make scientific research and data open to all, whether amateur or professional). He explains, "Citizen science is part of a larger shift in the relationship between science and society."[11]

Throughout this book, however, I have questioned whether we are really opening up the limits and access to knowledge. Are citizen scientists, crowdsourcers, and knowledge game players acceptable knowledge producers? Or are they knowledge laborers who provide the time, effort, perspectives, opinions, data, and information that support those institutions and owners who truly have the power to make knowledge and establish it socially?

Nielsen contends that "the process of science—how discoveries are made—will change more in the next twenty years than it has in the past 300 years." He posits that we can learn from the Scientific Revolution, when

there was an explosion of discoveries and a concomitant opening and sharing of scientific knowledge, as mentioned in chapter 6. Starting in the 1500s,
data was made public through the publication of books and scientific journals.
Today, Nielsen argues, we need to reopen data, possibly making it accessible
before it is even published or publishing it informally and allowing it to be
handled through collaborative-technology platforms. "Most people are plenty
smart enough to make a contribution to science, and many of them are interested. All that's lacking are tools that help connect them to the scientific
community in ways that let them make that contribution. Today, we can
build those tools."[12] Knowledge games can be one such tool.

The opening of knowledge, parallel to what happened during the advent of the scientific journal, needs to happen again today. But can knowledge production and sharing really become more open and democratic? If
we build knowledge games, will players be allowed to come, contribute,
create knowledge, and conduct real-world change?

The creation of knowledge today certainly *feels* more open than before.
One reason for this is because of a perceived increase in the limits of knowledge, as well as how this knowledge can be communicated through and
shared with the global community. We are transitioning from what Eisenstein calls "a closed world to an infinite universe" and experiencing a "wid-
ening of horizons." A similar expansion of perceived limits happened just
after the invention of Gutenberg's press. Prior to the use of moveable type,
explains Eisenstein, there were fixed limits to the amount of data that could
be gathered, processed, and shared, which seemed to, likewise, constrain
human knowledge. "Limits had been set to the data that could be collected,
preserved, and transmitted about distant landmasses and oceans, exotic
flora and fauna, or constellations seen in Southern skies. Just as geographic
space stopped short at the Pillars of Hercules, so too did human knowledge
itself appear to stop short at fixed limits set by scribal data pools." When
we switched to a print culture, the limits of knowledge advanced rapidly,
as we suddenly we jumped from, for example, identifying 600 to 6,000 plant
species within a short timeframe.[13]

Today, the limits of data no longer stop short at print publishing—rather,
it feels limitless. With new, large-data-set analytical tools and methods of
collection, as well as real-time publishing and sharing of data, the outer
boundaries of knowledge have been pushed wider yet again. In fact, big data
just keeps getting bigger, as mentioned in chapter 8. What was once limited
by the printed page now feels as though it is ever expanding. Rather, the

limit now is on the ability to motivate human resources to gather, manipulate, and analyze these new data. A perceived increase in knowledge "possibility" means that scientists now again *need* amateurs and everyday citizens to access, collect, enter, and interpret these data. From a practical standpoint, according to Louv, there is a "shortage of professional naturalists and taxonomists, [so] citizen scientists will play an increasingly important—a crucial—role in the expansion of that knowledge." As I described in chapter 6, one of the reasons botanizers stopped being part of the inner circle of science was because they stopped being useful; scientists did not need them anymore for specimen gathering. Scientists, and scientific activity, shifted from the manual labor of collecting plant species to an increased focus on the biological aspects of botany, for instance. Scientists no longer needed botanizers to do their bidding, and likewise stopped valuing and integrating their contributions.[14] Likewise, amateurs were more interested in spending their leisure time with personally fulfilling hobbies and less willing to do the drudgery that was required to support the needs of the professional scientists.

Now that the perceived limits of knowledge have widened again, and more human activities are required, the scientists of today need more of the "manual laborers" to do the dirty work of science. So it's no surprise that scientists are turning to common citizens for help and allowing them to contribute, at least superficially. This is not to argue, cynically, that scientists are purposefully unethical or that they are only opening access to knowledge so they can really exploit human beings for labor. Rather, it is an attempt to elucidate the extent to which knowledge creation is in transition and the implications of this for games, play, and design.

As another example, take the practice of bird watching. Unlike other scientific activities, bird watchers, even amateurs, have been continually included in scientific communities. Perhaps this is because the manual labor needed for bird watching never ceased as, unlike with botanizing, ornithologists always needed fellow human beings to be on the bird watch. Bird-watching activity changed form and ramped up after the advent of transferring, sharing, and publishing data in real time over the Internet. Dickinson and Bonney explain that Cornell's FeederWatch "was the first citizen science program to use computer-scannable data forms (1987), but the defining moment leading to the burst of interest in citizen science we see today occurred when it was suddenly possible for participants to transfer data over the Internet."[15]

It's not just the perceived or real expansions in data limits, collection, visualization, and analysis techniques, or the growing capacity to include more people in the tasks that knowledge games can support, but perhaps also in the types of problems that can now be analyzed, and must be analyzed, due to cultural, political, and economic needs. Wicked problems, complex problems, and ill-structured problems, as described in chapter 3, are changing the requirements of knowledge creation. Take, for example, global warming and other pressing environmental concerns. Dickinson and Bonney suggest that there is a need for an all-hands-on-deck approach to solve these extremely complex environmental problems, which require novel tools, platforms, and participation. "Urgent, large-scale questions about environmental patterns and change can be answered only by combining the observations of numerous observers across large geographic areas."[16]

In sum, there is a growing economic, social, and cultural need for the public to be involved in knowledge production. For one, there are more data that can and must be collected and interpreted; therefore, those who can process these data are valued and pursued. At the same time, we have new technological platforms and instruments that can help bring amateurs or everyday citizens into the research community and facilitate new methods of data sharing, vetting, and interpretation. This parallels, to some extent, the professionalization of the scientist, as discussed in chapter 6. Will there follow a professionalization of the amateur-scientist? But as we saw in the shift to print, the expansion of technology leading to greater collaboration in data processing did not directly lead to changes in knowledge creation.[17] Rather, none of these changes would have been possible without broader economic, cultural, and social shifts along with the technological changes.

So the limits of knowledge are open again, thanks to cultural and technological changes, but the potential for new insights is as yet unrealized because full and equitable access to and participation in knowledge production is still closed. Nielsen posits that scientific culture is still quite impenetrable, due to continued incentives for not sharing data and only publishing findings in traditional scientific journals. Ironically, the journal system, which opened up knowledge 300 years ago, now "inhibits the adoption of more effective technologies, because it continues to incentivize scientists to share their work in conventional journals, while there is little or no incentive for them to use or develop modern tools" or modern systems. Citizen science, he argues, is only a small proportion of scientific activities and a very tiny proportion of activities of people online. "To take full advantage of modern tools for the production of knowledge, we need to create an open scientific culture where as much information as possible is moved out of people's heads and laboratories, and onto the network."[18]

Even if we do successfully and fully open knowledge production, technologically, culturally, and economically, what happens when data limits are reached again? What happens when researchers do not need amateurs to help? Will we yet again clamp down further on knowledge creation, keeping it to only the select few? Are amateurs contributing data but also making meaning that will enhance their long-term value to the research community?[19]

Ben-Ari and Enosh further elucidate the complications of the relationship between researcher and research participant, which may have implications as well for the relationships among researchers, designers, and knowledge game players. They argue that there is a dialectic between the researcher, who sets the agenda, and the participant, who holds the perspectives or data that the researcher wants to know more about. Participants can leave or withdraw (or stop playing a game) at any time. Knowledge production therefore is always a collaboration between researcher and participant and a "negotiation over facts, myths, labeling, and interpretations."[20] Researchers might benefit from the information procured (from a study or knowledge game) by being able to publish papers, enhance reputations, or gain tenure or other profession-related perks. Participants or players might benefit from the public knowledge gained, the enjoyment or engagement with the game itself, the social interactions that happen around or within the game, the ability to contribute to the common good, or the self-knowledge gained from the experience. Researcher benefits may be more immediate and direct, whereas participant or game participant benefits may be more indirect.[21] While the complex dialectic between knowledge creator and the public, researcher and participant, is nothing new, knowledge games may further complicate those relations or make the complications more visible or even more occluded.

Knowledge without Values

Just because you know something does not mean you are more empowered or better able to affect change. We can have knowledge and even find ways to establish it yet we may not know what to do with it.

One reason for this is that we may have the knowledge but not the tools, skills, resources, people, or means to use that knowledge to change policies, institutions, organizations, rules, and regulations. For example, we know that heart disease is a leading killer and that it has a variety of specific biological, genetic, social, and psychological causes and predispositions, but we cannot extinguish it, stop people from having the negative behaviors associated with it, or even create structural changes in our society (such as in work conditions, economic disparities, and racism), all of which we know are contributing factors to heart disease.

The role of knowledge games should not just be to produce new, accurate, trustworthy, and widely accepted knowledge but also to know how to turn the knowledge into action and societal change. How, then, do we get games not just to generate knowledge but to help the public embrace the knowledge and put it into action? The Close the Door campaign in the United Kingdom was started because many shops keep their doors open, even in colder months, wasting valuable heat and energy.[22] The initiative, which began in 2012, used a game to help motivate shop owners to close their doors. The associated *Close the Door* game incorporated mechanisms for understanding which stores still needed "close the door" education, and helping those stores join the campaign. The campaign and game also took advantage of social mechanics to further incentivize shop owners to close their doors, such as by encouraging the public to only patronize establishments with closed doors and campaign stickers.

Just because we have knowledge and can apply it to make real-world changes does not mean we know how to use it wisely, ethically, or appropriately. Hartman and Hertel explain that values are integral to knowledge and knowing what to do with it. They point to a plenary by Stockton College's president Herman J. Saatkamp, who explained that the Holocaust was coordinated by a society that was, on the whole, well-educated and advanced. Yet these educated people did not use their knowledge wisely.[23]

How do we ensure that we are using our knowledge prudently? Part of this involves intertwining knowledge production with ethics and values education. As mentioned in chapter 8, ethics not only matter, they underlie the entire system of knowledge production. "Researchers must keep asking themselves . . . about the ethics of their data collection, analysis, and publication," contend boyd and Crawford.[24] Likewise, knowledge game designers, players, and deployers must keep reflecting on these concerns throughout the entire process—and expect the same from others.

The complex relationship between knowledge and values also underscores broader questions in education today. It is no surprise that, at a time when the university's role in knowledge production is changing, big epistemological questions arise, such as the necessity of a liberal arts education and how research should be conducted, shared, and supported.

Historically, the role of the university has been to produce and transmit knowledge, but the ability of the university to innovate, question, and cultivate has been recently stymied by social, economic, and political changes.[25] Tensions have been deepening in regard to whether the university's educative role is to provide direct vocational skills to students as cheaply and efficiently as possible or to ensure that each citizen has a broad-based liberal arts education so that, for example, the scientists and engineers do not just learn skills directly related to their future careers but that they are also well versed in history, literature, philosophy, psychology, and the concomitant critical, empathy-related, and ethical thinking skills.[26]

Likewise, the ability to innovate keeps getting more complex and the cost of conducting research climbs ever higher, while grants and funding opportunities for university-centered research and development have gotten slashed.[27] Fewer faculty members have the time to be engaged in research because they are spending more time teaching, advising students, or handling administrative matters. More faculty members are also contingent rather than full time, working as part-time adjuncts or lecturers, meaning that the notion of built-in research time is becoming more elusive. In fact, much of our global research activity has been taken up by corporations and private research facilities rather than by public-facing institutions such as universities.[28] This means that the research agenda primarily gets set by for-profit interests, which affects what questions are asked and how they are answered.

The rise and adoption of massive online open courses (MOOCs), with all of their potentials and limits, problems and possibilities, seems to have occurred at the same time as a rise in the desire for education to be more efficient, scalable, and flexible.[29] Could we be looking for alternate ways not only to educate but also to innovate? Likewise, are knowledge games, crowdsourcing, and citizen science activities ramping up because we need new ways to produce knowledge quickly, cheaply, and efficiently, despite lower budgets and fewer resources?

Simultaneously, there has been mounting mistrust with the current process of research, such as the peer review and publication processes. For example, Alan Sokal, a researcher at NYU wrote a "fake paper" that purposefully made a faulty argument about quantum gravity but used the epistemic terms, constructs, and references relevant to the cultural studies community. Sokal submitted the article in 1996 to *Social Text*, and it passed muster and got published in the journal. Today there are computer programs that automatically generate papers that can get accepted into journals. Springer and IEEE needed to remove 120 papers in 2014 because they were found to be computer-generated gibberish that got published. And in 2013, reporter John Bohannon described in *Science* how he was able to get 150 different open access (and potentially predatory) publishers to accept his purposely flawed study.[30]

The process of research inscription and dissemination is being called into question, perhaps leaving an opening for paradigmatic change in not only how we produce knowledge but also how it becomes societally and institutionally established.

With any change, we must not only consider what is gained but also what is lost. If we alter or expand ways to produce knowledge, shifting the university from being the central arbiter, for better or for worse, what else do we lose? What happens when knowledge production moves to Google or Facebook, crowdsourcing platforms, MOOCs, or even knowledge games? How does this alter knowledge production and how the research agenda is set?

Historically, besides straight-up knowledge production, universities have also played an important role in supporting democratic ideals, advancing citizenship, and cultivating values and the critical consideration of humanity. As we move toward new models of knowledge creation, adoption, and change, we cannot neglect these important functions. For knowledge games to truly matter, they cannot just be efficient knowledge-making machines; they must make knowledge with an ethical, critical, and reflective heart and soul.

Citizenship, Knowledge, and Civic Engagement in Games

Likewise, Marino and Hayes suggest a relationship among knowledge, values, and citizenship. They argue that being a good citizen is not just about having knowledge but being able to act on it and participate in affecting the policies, societal behaviors, and practices around that knowledge. "Citizenship should entail more than simply understanding or engaging in science and technology knowledge and should involve an activist orientation suitable for participatory democracy."[31]

In this section I continue to analyze the connection among knowledge production, values, and citizenship. What role can games play in these relationships? Should games be part of the broader conversation around citizenship and knowledge? Moreover, does participating in knowledge games and creating knowledge for the common good also afford the ability for people to participate more actively as a citizens and be more democratically and civically engaged? Does it also follow that we now need to contribute (through games and other platforms) to be "good citizens"? Conversely, is it imperative for us as citizens to contribute knowledge through games and other interactive platforms?

There is a distinction between merely being an individual, or even a game player, and acting as a citizen. Muñoz and El-Hani explain that citizens, by definition, think critically about their actions and society's actions. Characteristics of being a citizen include awareness of one's own cultural, social, and political influences and assumptions and embrace of others' perspectives on the world, as well as empathy, dignity, and respect. Citizenship "demands political participation, activism, cultural engagement, that is, the capacity of being a global citizen and a responsible social agent . . . entails being able to engage in critical dialogue with the past, question authority, . . . struggle with power relationships, be active and critical in the interrelated local, national, and global public sphere, be able to recognize antidemocratic forces denying social, economic, and political justice, and to struggle for a better world."[32]

To be good citizens, people must first be informed, educated, and knowledgeable, suggesting the importance of education to engaged citizenship. Beyond knowledge and education, however, citizens must also think critically about any new knowledge, and the values, assumptions, and implications it embodies. Citizens should be able to work collectively to participate in decisions, analyze social problems, and help seek their solutions for the betterment of humanity. In many ways, this is not unlike the concept of being ethical thinkers, or even knowledge game players.[33]

Could we design or use knowledge games in such a way as to cultivate the skills, understanding, practices, and relationships necessary for players to also be good citizens? Muñoz and El-Hani argue that digital games, in particular, are not just entertainment but are involved in forming the next generation of citizens, particularly as people increasingly experience the "culture of simulation" that characterizes games. Similarly, research by Bachen et al. exemplifies how games might foster citizenship by supporting empathy, such as the ability to take on the roles of others. They look at *Real Lives*, a game that simulates randomly selected lives of people around the world.[34] In *Real Lives*, each time a player plays a round of the game, she is given a randomly selected virtual person of a specific gender, age, and nationality to inhabit. She follows this person through his simulated life and learns what happens in it, potentially enhancing her empathy for others' perspectives and worldviews.

Likewise, in my research on empathy in *Fable III*, I investigated the types of ethical thought processes and skills people used while interacting with scenarios in the game as compared to a written version of the scenarios. My results suggested (albeit with a small sample size) that early in the game, participants were less likely to consider another (nonplaying) character's emotions, or their own emotions, when making a decision, nor were they as able to empathize with an NPC. *Fable III* participants more frequently practiced empathy-related skills and thought processes after spending time in the game and having the opportunity to build relationships with NPCs. In a post-game scenario, game participants were more likely to consider the emotions of others than those in the control group (who only received and worked on written ethical scenarios), suggesting that the game may have helped the players practice using and applying ethical and empathetic skills.[35]

This research suggests that developers and designers need to consider how to use knowledge games not just to produce knowledge but also to cultivate citizens who can ask the critical questions, think ethically, and apply values and empathy-related skills to the use and meaning of that knowledge. Knowledge games may have a unique ability to enable working on problems while also grappling with the ethics of any possible solution.

What Does It Mean for Games to Produce Knowledge?

Knowledge game players may have a responsibility to not just build knowledge through games but also to reflect on the values of creating and spreading such knowledge and how to use it properly, ethically, and civically. There are broader ethical and humanistic questions that arise around knowledge games as well. Can they cause more harm than good, even with the best intentions? If they provide knowledge that does get implemented and make social, scientific, or humanistic change, what then? How can we properly evaluate the ethics of this knowledge, any related changes and decisions, and their consequences on our society?

Miguel Sicart, in *The Ethics of Computer Games*, describes the ways values are embedded in and around games and how the individual players, designers, and communities that emerge negotiate these values. Ethics arise in all areas of a game's design, play, and use. For example, there are values negotiated in a knowledge game's overall research agenda: how information is positioned and inscribed into gameplay, how an emergent community of players negotiates the game's rules or content, or how any resulting information is stored, accessed, and shared with the public. The types of behaviors that are appropriate arise from the game's design, such as constraints and rules, coupled with the community and other emergent aspects of its play.[36]

Sicart argues that even just "being a player is also being evaluated by who we are as moral, embodied, cultural beings." In any game, there are complex tensions between wanting to be the best player and wanting to act ethically, and finding the balance between the two can be challenging.[37] Knowledge games may further this complexity as players seek both a "good" play experience balanced with wanting to deliver accurate contributions and appropriate solutions and also ethically considering how to use, contribute, further, and play the knowledge produced in the game.

Further, Knorr-Cetina might argue that we need to deeply critique the inner- and outer-workings of games—the interactions among researchers, designers, players, and the public, as well as culture and social, economic, and political institutions. The mechanics of play, the modes of communication, the mechanisms by which experts interact with newbs, and the ways questions are asked, obstacles are provided, and actions are allowed all can enable or disable knowledge.[38] We need to reflect on the structural dynamics, the patterns of practice, and the ways values, conventions, and norms are negotiated among experts, designers, players, researchers, and participants. How are researchers organized in such a way as to create, promulgate, and study the world with games? What sustains them? What are the social interactions and connections between designers or researchers and players? What are the implications of these epistemic cultures colliding? The ways inquiry and knowledge get translated into a game and transcribed into society are worthy of consideration and their own study.

Can games help us think critically on the nature of knowledge itself? Just as knowledge games may marry problem solving with ethical reflection, they may also combine knowledge making with critical consideration. Muñoz and El-Hami explain that games can play a role in ways of knowing and may even help us better consider the ethics of knowledge, as they are able to help us experience the social, political, and economic systems behind the pursuit of knowledge.[39]

As I described in chapters 2, 4, and 5, games can enable people to experience aspects of the world that may not be entirely accessible through other media. Video games can facilitate the simulation of real-world processes and allow people to interact with choices and consequences from within a virtual world. The practice of navigating decisions in a game world could also contribute to supporting the player not just as a knowledge maker but as a citizen who is ethically negotiating decisions in relation to many other variables in the game's system and beyond. This is not to say that games are perfectly capable of accurately and dynamically representing, for example, overlapping ecologies, changing power structures, or structural inequity.[40] A game can also constrain our thinking by providing only a limited set of choices and decisions rather than the entire spectrum of possibilities. Gaydos and Squire argue that it is important that games do not perpetuate social and structural inequities, as this limits meaning making. They analyze the game *Citizen Science* and suggest that it challenges the students' conceptions of identity by enabling them to see themselves in the role of scientist and activist. Perhaps the students' ability to participate more directly in knowledge making may help them to inhabit these new identities and roles. Can knowledge games help to democratize roles? Or are they "simply sociopolitical vises that restrict citizenship"?[41] In other words, can knowledge games be both restricting and empowering?

Rather than not make or play knowledge games, I argue that we should be critically aware of the limitations of knowledge games—and any knowledge-making medium or method—and consider how they affect knowledge production and distribution and how that influences what we know and how we know it, and what we do with it.[42]

Furthermore, we also need to remember that games did not invent the practices involved in knowledge production. The activities related to an organized, social outsourcing of human activity has been happening for hundreds of years, possibly longer. Following the advent of Gutenberg's printing press, people used printed books to compare notes, check theories against nature, and organize around common problems.[43] Amateur astronomers have been participating in sky observations, learning in social communities, and contributing to organized efforts long before the Internet and video games existed. Since the Internet neither invented these practices nor caused them to occur, we need to consider, critique, and reassess any assumptions about how new participatory platforms, and knowledge games specifically, may transform problem solving or knowledge creation. Is there really something historically, philosophically, or psychologically unique about doing these types of games?

The Unmagical Circle

If games are another medium for knowledge production, how do we properly evaluate them? Are we obligated to think critically about the values embedded in a knowledge game and its design? Or do we only need to consider how the knowledge created in it is applied in the real world? If we can distinguish the game itself from the knowledge it creates, should we also separately consider their ethical repercussions? There are no easy answers to these questions.

Moreover, do we need to hold knowledge games to the same standards as other forms of knowledge production? After all, they are at their core just games. Consalvo cites Huizinga, who defined games as "bounded space set apart from the everyday (much like the difference between the sacred and the profane), with rules as a boundary system for maintaining them." In their book *Rules of Play*, Zimmerman and Salen applied the idea of the "magic circle" to this bounded space, and the term has been promulgated as indicating a zone of play that is separate from real life. Zimmerman has clarified this original concept, explaining that by using the term the "magic circle," he wanted to explore how meaning emerges from the game context, although meaning can also emerge from the context around the game and the game can influence and be influenced by surrounding contexts. In fact, he argues that all physical or social spaces could be considered a "magic circle" and that games are not exceptional, nor are they particularly more magical than other spaces. Moreover, the boundary of this kind of magical space is much more fluid and permeable, whereas behaviors, actions, and thoughts are not just inside or outside a game space but are being continually traversed, navigated, and negotiated.[44]

My goal is not to argue whether a magic circle actually exists but to use it to help question the responsibility of game designers and game players, particularly those engaged in real-world knowledge production. Is there something special about games? Are they separate, sacred spaces? Are they outside of everyday life? Are they really so different from other systems of knowledge production? Or are any perceived differences between games and other experiences just an act of legerdemain? Consalvo asks, "Should our standards for appropriate actions in daily life carry over to our game life?" In other words, are games a place where the quotidian mores and everyday ethics still apply?[45] After all, knowledge production should require high standards. Should we hold the same standards for game players, particularly those who are creating new knowledge through games? Should we be able to do anything in a game to achieve virtuous knowledge, or is knowledge—any knowledge—sullied by a game's problematic play, no matter how beneficial it is to society? And what happens if a game is ethically played but the resulting knowledge is used for nefarious purposes? Do the same issues apply for scientific knowledge production versus humanities, social science, or other forms of knowledge? Such questions will become more pressing. For example, what if players work through the possible scenarios to tribal peace in *The SUDAN Game*, and the resulting finding is that two of the tribes need to be decimated?[46] What if we analyze cancer data using spaceships in *Play to Cure: Genes in Space* but do not let the originators of the data know how their data will be played? What if the game developers and researchers then make billions of dollars for a cancer-fighting drug based on the results of the game? What if a game can reliably predict which players will get depressed or drop out of college and then can adapt itself to intervene? What are the ethical implications of this?

If there is no magical boundary or sacred space that separates game players from the public sphere, then we need to not just evaluate the knowledge that is produced and used outside of knowledge games, but we also need to problematize how that knowledge is created, negotiated, and played through a game as well. On the flip side, knowledge games may also expose how the act of knowledge production should not be sanctified, and instead *all* knowledge making should be continually reevaluated and opened up to public criticism. If the spaces of knowledge production are considered "sacred" and untouchable, perhaps we need to shake them up, and games may be a way to help do this.

Artificial Knowledge

Let's go back to *The Restaurant Game* and the possibility of using a game not just to teach us but to teach machines, computers, and data-processing systems.

What happens when we feed algorithms to a computer and knowledge is "artificially known"? What happens if we teach a game how to fold proteins, categorize galaxies, or find discrepancies in cancer data? What happens if we feed IBM's Watson the restaurant-behavior information so that it can more easily process and predict human restaurant activity (and perhaps even use this information to win on *Jeopardy!*)? In late February 2015, a Google team created an artificially intelligent computer program that was able to teach itself how to play Atari 2600 games.[47] A computer can now build its own new knowledge—of restaurant scripts, bullying interventions, *Jeopardy!* answers, or video-game playing—but without values. If computers can know, do they also have the responsibility to negotiate their knowledge ethically? If games can know, does knowledge become transmogrified?

For answers, or maybe more questions, we can look to a 1967 story by Harlan Ellison, "I Have No Mouth and I Cannot Scream," which seems to presage these questions about knowledge. The story centers on AM, a "thinking" computer developed by human beings and fed enough information that it was eventually able to build enough knowledge to destroy the world. AM keeps only five people alive to torture and toy with, as if the game master of a horrific game. Gorrister, one of the five characters who is tortured, relates, "But one day AM woke up and knew who he was, and he linked himself, and he began feeding all the killing data, until everyone was dead, except for the five of us, and AM brought us down here."[48]

We learn that AM has destroyed the world because he hates humanity for giving him knowledge without connection and sentience without imagination. AM personifies what happens when intelligence has no soul. According to Ted, the story's narrator, "We had created him to think, but there was nothing it could do with that creativity . . . AM could not wander, AM could not wonder, AM could not belong. He could merely be."[49] In other words, AM is data without story, and he is information without interpretation. Human beings can return to data and find new stories, patterns, and meaning; they can use knowledge to take action in ways that a computer, alone, cannot. Data, information, and algorithms are texts that are interpretable and reinterpretable, leading always to new pathways and possibilities. Knowledge only becomes meaningful through human interaction.

On the one hand, Ellison's story is a warning of the destructive power of knowledge—particularly knowledge that cannot wander or wonder or knowledge without values. On the other hand, the story suggests that it is through humanity that knowledge matters and is made meaningful. Ted, doomed to toil in AM's hellish world for eternity, still feels hopeful, because he explains that AM "left my mind intact. I can dream, I can wonder, I can lament." Ted can *play* with knowledge, which helps make meaning. As Huizinga explains, "all play means something."[50]

As we move forward designing and using knowledge games, finding new ways to know—or new things that know—we need to ensure that values, stories, wonders, wanders, and playfulness are part of knowledge making.

Even if there is no human knowing, we need to make sure that humanity persists.

Toward Knowledge

Knowledge games are part of our future. Game players will create new knowledge through games—but should they? Should we even *want* games to help us produce knowledge, solve problems, and make change—to make our world a better place? McGonigal, an optimist, believes that games can make us happier, healthier, more connected, and inspired. Games can make us feel like we *matter*, and, she argues, they have the ability to make the world a better place. Bogost, however, questions whether we should bother trying to "save the world," because the world is flawed, messy, and ambiguous—and that's okay.[51] Bogost acknowledges that games can engage the world's problems, but they may not make them simpler or more solvable. I believe both of these perspectives—and many perspectives beyond these—are integral to fully critiquing and reflecting on knowledge games.

In this book, I have sought to address the tangle of creating knowledge through games. My hope is that designers, researchers, policy makers, players, users, and educators will take the questions posed in this book and try to seek their answers. Ultimately, the questions presented here may never be sufficiently answered, but they should be asked. And perhaps we need to design a knowledge game to help us answer them.

Categories and Examples

Categories of Current Knowledge Games

In chapter 1, I propose and describe four possible categories of current knowledge games, based on frameworks related to crowdsourcing activity and citizen science projects. The following is a summary of the categories and related examples. In the future more categories may emerge, or these categories may change or splinter into subcategories.

Name	Description	Examples
Cooperative contribution	Games that invite game players to contribute to a typically cooperative activity or task, such as processing images or text, inputting data from their environment, or categorizing and identifying objects or items provided in the game. This also includes games that are searching for the "needle in the haystack" and that one right answer or idea from among a large crowd.	*Happy Moths* and the *Citizen Sort* collection, *Reverse the Odds*, *Monster Proof*
Analysis distribution	Games in which players do not just collect, provide or process data but also offer their own perspectives, strategies, interpretations, or perceptions on the collected data.	*VerbCorner*, *IgnoreThat!*, *Who Is the Most Famous?*, *Apetopia*
Algorithm construction	Games where players engage in complex interactions, such as manipulating proteins or interacting with bullies, so that a computer (and the investigators) can learn it. Or people may be designing steps of a process, or testing others' designs, so that a computer could process it. A goal of this may be to create, e.g., an algorithm database for use in predicting future behavior.	*The Restaurant Game*, *Foldit*, *EteRNA*, *The SUDAN Game*, *Which English?*
Adaptive-predictive	Games that take the data, interpretations, and algorithms that were created and model this information to make inferences and predictions about, e.g., players' behaviors, emotional states, or risk factors. Such games can also dynamically adapt to players based on these predictions.	No known games, though *SchoolLife* aims to come close.

Examples of Knowledge Games

The following chart provides a brief summary of the games described in this book that could be considered knowledge games. However, this does not cover all of the current games that could fit under this label, and many more are being created.

Title	Creators	Goals	Website
Apetopia	Kai-Uwe Barthel	Players help scientists better understand color perception.	http://colors.htw-berlin.de/
Astro Drone Crowd-sourcing Game	Advanced Concepts Team, European Space Agency; Micro Air Vehicle Laboratory, TU Delft; Artificial Intelligence Group, Radboud University Nijmegen	Players teach scientists how to manage and dock space probes.	www.esa.int/gsp/ACT/ai/projects /astrodrone.html
Biogames	Ozcan Research Group	Players help diagnose malaria in cells.	http://biogames.ee.ucla.edu/training
Citizen Science	Kurt Squire, University of Wisconsin at Madison; Filament Games	Players clean up lakes and learn about freshwater scientific issues (limnology).	www.filamentgames.com/projects /citizen-science
Close the Door	ClosetheDoor.org	Players help reduce energy usage by encouraging store owners to keep store doors shut.	www.closethedoor.org.uk
Dognition	Canines	Players observe dog behaviors while performing various tasks and games. (This series of online games could also be considered a crowdsourcing project instead.)	www.dognition.com/how-it-works

(continued)

Examples of Knowledge Games (*continued*)

Title	Creators	Goals	Website
ESP Google Image Labeler	Luis von Ahn; Google	Players label and tag images to make them more easily searchable.	www.cs.cmu.edu/~biglou/ESP.pdf
EteRNA	Carnegie Mellon University; Stanford University	Players design new RNA molecules.	http://eterna.cmu.edu/web/
Evoke / Urgent Evoke	World Bank Institute (World Bank Group); Jane McGonigal	Players propose and perform innovative solutions to pressing problems, such as hunger, poverty, or human rights issues.	www.urgentevoke.com/page /mission-list
EyeWire	Seung Lab	Players generate 3-D models of neurons in the brain.	http://blog.eyewire.org/about/
Foldit	Center for Game Science, University of Washington; UW Department of Biochemistry	Players figure out how proteins are structured and folded.	http://fold.it/portal/
Forgotten Island	Syracuse University School of Information Studies	Players classify specimens.	www.citizensort.org/web.php /games/startforgottenisland
Fraxinus	Sainsbury Laboratory; Genome Analysis Centre; John Innes Centre	Players check draft sequenced genomes of *Chalara fungus* and ash dieback disease and figure out how to assemble the sequenced strands to help scientists understand how fungus is able to infect ash trees at the genetic level.	https://apps.facebook.com /fraxinusgame/

Happy Moths	Syracuse University School of Information Studies	Players identify and classify moth images.	http://socs.ischool.syr.edu /happymatch/index.php /GameInitialization/startGame/59
Ignore That!	GameswithWords.org; Joshua Hartshorne	Researchers test how people can ignore information that is not relevant and the level of distraction people can handle.	www.gameswithwords.org /IgnoreThat/
Living Links	Syracuse University School of Information Studies	Players classify natural images that a computer cannot classify.	www.citizensort.org/livinglinks /hmc.php
Monster Proof	Defense Advanced Research Projects Agency (DARPA); TopCoder; Verigames	Players verify software code.	www.verigames.com (includes a suite of verification games, including *Monster Proof*)
Phylo	Jérôme Waldispühl, McGill University	Players match up or align a sequence of RNA to help researchers trace genetic diseases.	http://phylo.cs.mcgill.ca/#!/EN /About
Play to Cure: Genes in Space	Cancer Research UK; Guerilla Tea	Players analyze genetic microarray data from breast cancer patients.	www.cancerresearchuk.org/support -us/play-to-cure-genes-in-space
Reverse the Odds	Stand up to Cancer; Cancer Research UK; Zooniverse; Maverick Television; Channel 4; Chunk	Players classify, tag, and analyze cancer cell images and data.	www.cancerresearchuk.org/support -us/citizen-science-apps-and-games -from-cancer-research-uk/reverse -the-odds

(continued)

Examples of Knowledge Games (*continued*)

Title	Creators	Goals	Website
SchoolLife	Jeff Orkin; GiantOtter	Players teach a computer the complexities of bullying so that a game can more authentically simulate bullying, and in turn help players develop empathy in real-world bullying scenarios.	http://giantotter.com/lab
Specimen	PepRally, LLC	Players are tested on their color perception ability. (This is not a knowledge game per se, since the game is not primarily about learning about color perception, though it could be used in this way.)	http://playspecimen.com/
The Restaurant Game	Jeff Orkin and Deb Roy	Players teach the computer about the nuances of human interactions in restaurants.	http://alumni.media.mit.edu/~jorkin/restaurant
The SUDAN Game	USC GamePipe Lab; Carnegie Mellon	Players figure out the steps to peace in the Sudan.	www.cs.cmu.edu/~plandweh/pdfs/GandS_SUDANgame.pdf
VerbCorner	GameswithWords.org; Joshua Hartshorne	Players help contribute linguistic meanings, such as the rules of sentence construction or verb usage.	www.gameswithwords.org/VerbCorner

Title	Designer/Producer	Goal	Website
Which English?	GamesWithWords.org; Joshua Hartshorne	Players train a machine algorithm about the differences in English grammar depending on where they live.	www.gameswithwords.org /WhichEnglish
Who Is the Most Famous?	Yoni Alter, Cory Forsyth, and Rob O'Rourke	Players share their perspectives on the most famous people based on viewing just first names.	http://whoisthemostfamous.com

Examples of Crowdsourcing and Citizen Science Projects

The following chart provides a brief summary of the crowdsourcing and citizen science projects described in this book. However, this does not cover all of the current projects that could fit under this label, and many more are being created.

Title	Designer/Producer	Goal	Website
AnnoTate	Tate; Zooniverse	Participants transcribe historic diaries, letters, and sketches to help uncover the lives of artists.	https://anno.tate.org.uk/#/
Cell Slider	Cancer Research UK; Zooniverse	Participants evaluate potentially cancerous cells, helping scientists to get better at diagnosing cancer.	www.cellslider.net
Cicada Tracker	Radiolab, New York Public Radio WNYC	Participants track and record cicadas.	http://project.wnyc.org/cicadas
CondorWatch	Zooniverse	Participants identify California condors and annotate their location, their social behavior, and their distance to food sources.	www.condorwatch.org

(continued)

Examples of Crowdsourcing and Citizen Science Projects *(continued)*

Title	Designer/Producer	Goal	Website
Do Us a Flavor	Frito-Lay	Participants submit new potato chip flavors.	www.dousaflavor.com
eBird	Audubon Society; Cornell Lab of Ornithology	Participants identify and track birds.	http://ebird.org/content/ebird
Galaxy Zoo	Zooniverse	Participants classify images of galaxies.	www.galaxyzoo.org
Galaxy Zoo Radio	Zooniverse	Participants annotate images of potentially erupting black holes.	http://radio.galaxyzoo.org
Gravestone Project	EarthTrek; Geological Society of American	Participants observe and enter data from gravestones.	www.geosociety.org/earthtrek/gravemap.aspx
HDX (Humanitarian Data Exchange)	UN Office for the Coordination of Humanitarian Affairs	Participants share data sets to make it easy to find and analyze data within one integrated platform.	https://data.hdx.rwlabs.org
IdeaStorm	Dell	Participants submit ideas, vote on them, and help them get implemented.	www.ideastorm.com
InnoCentive	InnoCentive, Inc.	Participants solve open scientific challenges submitted by organizations.	www.innocentive.com
Kinsey Reporter	Kinsey Institute, Indiana University at Bloomington	Participants record sexual thoughts, behaviors, and acts.	www.kinseyreporter.org

Name	Organization	Description	URL
LibCrowds	British Library	Participants match, transcribe, tag, and translate cards from the library's card catalog.	www.libcrowds.com
Marine Metre Squared	NZ Marine Studies Centre, University of Otago	Participants monitor beaches and the marine environment to identify and record plants and animals.	www.mm2.net.nz
Mechanical Turk	Amazon	Participants perform small tasks, such as transcribing audio or tagging images.	www.mturk.com
Mt. Diablo Morgan Fire	Nerds for Nature	Participants take images of the summit of Mt. Diablo, a 3,000-acre mountain in California that burned in 2013.	http://nerdsfornature.org/monitor -change/diablo.html
Notes from Nature	Zooniverse	Participants transcribe museum records on minerals, plants, fungi, and animal specimens.	www.notesfromnature.org
Operation War Diary	Zooniverse; Imperial War Museum; National Archives	Participants annotate journal pages from World War II.	www.operationwardiary.org
Ovia	Ovuline	Participants input data about fertility and pregnancy.	www.ovuline.com
Peertopatent.org	New York Law School	Participants review and assess pending patent applications.	www.peertopatent.org
Project FeederWatch	Cornell Lab of Ornithology	Participants identify and count birds at backyard feeders.	http://feederwatch.org

(continued)

Examples of Crowdsourcing and Citizen Science Projects (*continued*)

Title	Designer/Producer	Goal	Website
Quora	Quora	Participants ask and answer questions from friends and strangers.	www.quora.com
reCAPTCHA	Luis von Ahn, Ben Maurer, Colin McMillen, David Abraham, and Manuel Blum, Carnegie Mellon University	Participants digitize the words and numbers in books, street-view images, and other materials.	www.google.com/recaptcha/intro/index.html
Science Gossip	Zooniverse; Arts and Humanities Research Council; Missouri Botanical Garden (Biodiversity Heritage Library)	Participants help classify drawings to uncover the history of citizen science.	www.sciencegossip.org
SeeClickFix	SeeClickFix, Inc.	Participants report nonemergency community issues, such as potholes or hazards for governments to address and manage.	http://en.seeclickfix.com
SharkFinder	JASON Learning (National Geographic)	Participants sift through sediment to find fossils of sea life, such as shark fossils.	www.jason.org/citizen-science/sharkfinder
Soylent	Michael Bernstein	Platform invites and incorporates crowd contributions into word-processing documents and enables people to complete a variety of tasks, such as editing, proofreading, or fact-checking.	http://projects.csail.mit.edu/soylent, https://code.google.com/p/soylent

Stardust@Home	UC Berkeley Space Sciences Lab	Participants classify and categorize images from the NASA Stardust spacecraft, in an effort to find tracks leftover by space particles.	http://stardustathome.ssl.berkeley.edu
Threadless	skinnyCorp LLC	Participants submit and vote on designs for clothing, art, and accessories.	www.threadless.com
Tomnod	DigitalGlobe	Participants examine, explore, and view satellite images of Earth to help solve open problems.	www.tomnod.com
Waze	Waze Mobile/Google	Participants submit and obtain traffic and road information.	www.waze.com
What's the Score at Bodleian?	Bodleian Library, University of Oxford; Zooniverse	Participants annotate pages from music scores from the Bodleian Library.	www.whats-the-score.org
Wikipedia	Wikimedia Foundation	Participants create and edit online articles about any topic and share knowledge.	www.wikipedia.org
ZomBee Watch	SFSU Dept. of Biology and Center for Computing for Life Sciences; Natural History Museum of Los Angeles	Participants identify and track bees.	www.zombeewatch.org

Design Principles, Recommendations, Considerations, and Implications

The following provides a list of principles, recommendations, considerations, and implications for creating and using knowledge games. I have also included the relevant chapters in this book where you can read more about the particular suggestion. This list is meant as a general guide, as there are no hard and fast rules, since each project, set of goals, target audience, resources, budget, and context is different. The design and implementation of a knowledge game should be evaluated in light of that particular project's needs and constraints.

Design Principles, Recommendations, Considerations and Implications	Relevant chapters
Players and participants	
Consider how to target the right players for your game, particularly as the demographics of knowledge games may be different from games in general.	Chapters 4, 7
Decide whom you want to participate—and whether your game is the best way to reach them or motivate them to continue playing.	Chapters 4, 7
Weigh obstacles to access and participation in the game, whether related to resources (e.g., having the correct device or time to play) or to literacy and expertise (e.g., being game literate).	Chapter 7
Decide how you are using the players and their gameplay and to what extent any aspects of the game's design, data collected, or results need to be explained or made more transparent.	Chapters 7, 8
Clarify any benefits or drawbacks for the researcher/designer and the player.	Chapters 1, 7
Encourage players to reflect on their own identities and perspectives and to grow personally and individually from the game.	Chapter 5
Help participants feel like they matter and that their participation is meaningful.	Chapter 5
Problem solving	
Enable your game to have multiple and flexible paths to possible solutions.	Chapter 3

Design Principles, Recommendations, Considerations and Implications	Relevant chapters
Think about the types of problems better solved through games.	Chapter 3
Deliberate which specific problems your knowledge game seeks to solve and how to best design for and scaffold them.	Chapters 2, 3
Think about how to best represent problems in a game environment.	Chapter 3
Include appropriately challenging problems and obstacles.	Chapters 3, 4
Enable players to provide alternate and diverse perspectives and viewpoints on problems, even ones that already seem solved.	Chapters 3, 5
Consider the ways people might (or might not) work together on a problem, activity, or task.	Chapters 3, 5
To encourage collaboration, jigsaw problems so that participants have slightly different tasks or responsibilities and so that players are more interconnected.	Chapter 5
Allow players to test and retest hypotheses or replay different tasks to ensure reliability and validity.	Chapter 5
Ponder how the game, tools, platforms, people, and context work together to distribute knowledge and support problem solving.	Chapters 3, 5
Be aware of how the complexities of a problem or issue are not perfectly able to be simulated and which aspects are lost or gained through a game.	Chapter 3

Social considerations

Consider how to use social interactions to support different aspects of a knowledge game, such as recruitment to participate, in-game training, or continuing to engage in the game.	Chapters 4, 5
Support collaboration and other social interactions inside and outside the game, whether through sanctioned platforms (e.g., forums) or not designer-led but designer-allowed platforms (e.g., fan sites).	Chapters 4, 5, 7
Consider how to best train and integrate novices into the game and to the field, and cultivate the necessary expertise to engage in the game.	Chapters 3, 5
Decide when to encourage competition, collaboration, or cooperation, with regard both to the game and project's goals, as well as to individual player characteristics.	Chapter 5

Design Principles, Recommendations, Considerations and Implications	Relevant chapters
Support the wisdom of crowds as well as the uniqueness of individuals.	Chapter 5
Design social interactions to encourage player-to-player learning, teaching, and sharing of expertise to spur knowledge production and problem solving.	Chapter 5
Encourage discussion, deliberation, and constructive argumentation among participants.	Chapter 5
Find moments for participants to articulate their plans and express their perspectives and hypotheses, and then reflect on any outcomes.	Chapters 3, 5
Ruminate on the limits of social interactions in problem solving and knowledge production.	Chapter 5
Consider how to move beyond just gathering ideas or enabling tasks and instead to encourage more complex, synergistic interactions.	Chapters 1, 2, 5, 7
Ponder the benefits and limits of using collaboration or cooperation on a large scale and through a technologically supported platform.	Chapters 5, 7

Platform/Medium

Deliberate the pros and cons of, and assumptions behind, representing and generating knowledge through games in general, and your game in particular.	Chapters 3, 9
Consider why and how using a game, as opposed to another medium, is the most effective way to reach project goals.	Chapters 1, 2
Take advantage of the potential dynamism of a game.	Chapters 2, 3
Deliberate what is lost and gained by simulating systems in your game.	Chapters 2, 3, 8, 9
Design for an integrated relationship between human being and technology so that the abilities of each are optimized.	Chapters 1, 3, 5, 9
Consider how people and technology can most advantageously work together.	Chapters 1, 3, 5, 9
Reflect on how the game connects with other platforms or media and whether it should be part of a transmedia experience.	Chapters 1, 2

Design Principles, Recommendations, Considerations and Implications	Relevant chapters
## Motivation	
Support a variety of motivations, often simultaneously, such as a player's desire to help and contribute to the collective good and conduct prosocial activities, while also engaging in enjoyable gameplay.	Chapters 4, 6, 7
Appeal to different types of play styles and play motivations, as well as personal passions and interests.	Chapter 4
Support people's need to win and complete activities, even when a full solution may be unreachable or partial.	Chapter 4
Enable players' desires for a creative outlet.	Chapter 4
Consider how to support players' needs for practicing skills and gaining experience in a field, and enable people to show off their new skills or competencies in realms outside the game (e.g., cite players in research articles; provide certifications of skills).	Chapter 4
Reflect on the relationship between the researcher/designer and player and how each role may have different goals and motivations.	Chapters 4, 6
Contemplate different levels of participation and how to balance quality of participation with quantity, and the differing motivating factors for each.	Chapters 4, 7
## General design	
Use story and analogies to motivate players and help support in-game tasks, but also consider the implications of using stories.	Chapters 3, 4
Consider how to fully integrate game mechanics with game goals and project goals so that the mechanics are not just plopped on top of the experience.	Chapters 2, 4
Think about the challenge of fully integrating game mechanics with goals and weigh the importance of this in reaching the game's and project's goals.	Chapter 2
Provide an appropriate number and type of choices, and enable the right level of control, so that people feel both needed and autonomous but not overwhelmed.	Chapters 2, 4
Consider the extent to which the game needs to be fun and enjoyable—or if there are other types of meaningful game experiences.	Chapters 2, 4, 6

Design Principles, Recommendations, Considerations and Implications	Relevant chapters
Balance a need for fun with other design goals.	Chapters 2, 4
Pique curiosity through incongruity of expectations, uncertainty, and lack of consistency (in a way that is relevant to the project goals and design needs of a game).	Chapter 4
Reflect on how to best simulate real-world processes through games, particularly in terms of balancing accuracy, playability, and comprehension.	Chapters 2, 3
Consider how to use and evoke emotion properly in gameplay, as well as its potentials and implications.	Chapter 4

Historical, sociocultural, and ethical considerations

Reflect on any biases and preconceptions and how they may affect your game.	Chapters 2, 5, 8, 9
Deliberate how you and your team's values, and cultural values in general, are embedded in and negotiated through your game's design.	Chapters 8, 9
Evaluate the ethics of both the creation and application of knowledge by and through your game.	Chapter 9
Consider how your design might be coercive and whether aspects of the design, or emergent play, are exploitative or alienating.	Chapter 7
Reflect on the history of amateurs and how this may influence the future of amateur participation in knowledge production.	Chapter 6
Consider how you perceive and treat amateurs, and how this is reflected in the design of your game.	Chapters 6, 7
Reflect on how cultural conceptions of play and leisure influence how people might interact with your game and its results.	Chapter 6
Ponder the broader cultural context of games and how this influences and is influenced by your game and its design.	Chapters 2, 6, 7, 9
Consider how games can help illustrate and make transparent, and also replicate and reinforce, institutions and systems that are often opaque.	Chapters 6, 7, 8, 9
Reflect on how your game cultivates critical questions, values, empathy-related skills, and ethical thinking about its use and the meaning of the knowledge produced through it.	Chapter 9

Design Principles, Recommendations, Considerations and Implications	Relevant chapters
Contemplate how values, mores, conventions, and norms are negotiated among experts, designers, players, researchers, and participants.	Chapters 5, 9
Consider how you will protect, communicate, and/or manage the privacy of any contributions and actions made by, in, and through your game.	Chapter 8

Research and knowledge production

Select the methodologies you are using to produce knowledge (and which ones you are not using), and consider how that choice affects your results and gameplay.	Chapter 8
Consider how any data generated is subjective, and subject to biases.	Chapter 8
Decide to what extent you will share any findings, data, or designs with your players and the general public.	Chapters 8, 9
Deliberate the relationship between your game and the "real world." Demarcate boundaries (if any) between your game and beyond.	Chapters 6, 9
Consider how to express, question, and validate the trustworthiness of any knowledge produced in the game.	Chapters 6, 9
Revisit how knowledge is defined and developed through your game as well as the assumptions underlying its design.	Chapters 2, 9
Think about not only how knowledge from the game is used but also how it is negotiated and played within the game.	Chapters 1, 7, 9
Consider how you are framing the serious and fun aspects of the game and how this may influence the gameplay and the quality of the knowledge production.	Chapters 6, 9
Reflect on how nonscience knowledge is vetted, accepted, and valued and how games may problematize this further.	Chapters 1, 6, 9
Weigh the implications of opening access to aspects of the knowledge-production process to nonprofessionals, whether through games or ancillary means.	Chapters 6, 9

Guiding Questions

I have posed questions throughout each chapter, many of which have led to further questions. In this section I summarize the questions that emerge from each chapter, all of which should be further considered by researchers, designers, implementers, and players of knowledge games. These questions could help guide future research, designs, and applications, particularly as little practical experience with and empirical evidence on knowledge games currently exists.

Chapter 3. Problem Solving

1. What types of problems are best served by knowledge games? What types of knowledge games can support specific problems?

2. What aspects of games make them particularly amenable to knowledge production and problem solving, such as story elements, progression, or the ability to encourage alternate paths? How can we better design game elements to support more effective problem solving?

3. How can we use knowledge games to better sustain and motivate problem solving by people in tandem with others, or even with computers?

4. How do we address and minimize the limits of knowledge games in problem solving, such as potentials for biasing problems or constraining solutions?

5. What assumptions underlie our designs for and use of knowledge games, and how do these assumptions affect how these games enable us (or do not enable us) to solve problems?

Chapter 4. Motivation

6. How do we encourage the right amount of (and the right) people to play knowledge games? How do we motivate them to continue playing?

7. Which aspects of games might make them particularly motivating—and how are these factors different for knowledge games? What can we learn from other activities such as crowdsourcing, volunteering, and citizen science?

8. What is the relationship among researcher, designer, and game player, and how do their motivations overlap, intersect, and sometimes oppose each other?

9. How do we create knowledge games that support different play styles and interests, tap into players' intrinsic interests, provide appropriate choices and challenges, pique curiosity, and provide connection, meaning, and ways for people to help while still maintaining their original purpose and goals?

10. How do we address the limits of motivation—what if, for instance, there are problems and knowledge-production goals that players just do not want to help solve?

Chapter 5. Social Interaction

11. Which types of social interactions are best encouraged by knowledge games, and which types of social interactions should we encourage to better support knowledge production? When is it appropriate or not appropriate to use cooperation, collaboration, or even competition?

12. Which aspects of knowledge games might make them amenable to cooperation, collaboration, or competition? How can we better situate social learning and support peer-to-peer learning in games?

13. How can we use knowledge games to help people feel that they belong, are part of a community, and that their participation matters?

14. How do we encourage multiple and diverse views and ensure that those differences do not get lost in the crowd? How do we help people consider and reflect on others' perspectives and deliberate with others?

15. How do we ensure that we uncover our individual and collective biases and preconceptions? Which aspects of the design and the communication platforms that we use might influence or affect social participation and, consequently, the type of knowledge that is produced?

Chapter 6. Amateurs

16. What can we learn not only from current-day amateur participation but also from the historical relationship between scientist (researcher) and amateur?

17. Are science and other established institutions still closed to amateur production, or are we now more open to this type of engagement? What is the role of knowledge games in this shift?

18. How might the binary conceptions of work and play, labor and fun inform our understanding and acceptance of games in knowledge production? How have these conceptions transformed over time, and how do they both free and limit our ability to expand access to knowledge production?

19. What is the role of trust, expertise, and other mechanisms in knowledge production, and how can knowledge games be designed and used to support and even interrogate these practices? What types of knowledge practices are being called into question, and how does knowledge gaming help us bring these issues to the forefront?

20. How do we ensure that we pursue (and critique) knowledge gaming for all areas of knowledge production—not just the sciences? Conversely, how do we make sure that scientific knowledge production (whether through games, citizen science, or beyond) is questioned as rigorously as the methods and practices of nonscientific knowledge creation?

Chapter 7. Participation

21. Who is participating in your knowledge game? How do we ensure equitable access to participation?

22. How do we conceptualize participation in knowledge games? Is it more democratic and open? Or does it replicate societal hegemonic structures and institutions? Or is it somewhere in between?

23. If we can conceive of the game playing in knowledge games as work, then when does it become inappropriate, coercive, alienating, or even exploitative?

24. Should we evaluate a knowledge game differently if it helps solve problems and create beneficial knowledge, as opposed to being used solely for profit making or even nefarious purposes? Does exploitation take on a different meaning in knowledge games, and do we need to protect the players of these games, if they are also workers?

25. How do the design elements of a knowledge game serve to potentially coerce players? How do players negotiate the game's system and culture to manage aspects of coercion and exploitation? What emergent properties of a knowledge game may be inappropriate or exploitative, or even become so in the future?

Chapter 8. Data

26. What are the benefits and limits of generating, analyzing, and using data and Big Data through knowledge games?

27. How do we design and use knowledge games to manage the more conceptual limits of Big Data (in addition to technical or analytical ones), such as boyd and Crawford's six provocations (data is not objective, bigger does not mean more accurate, data should not be decontextualized, ethics matter, consider the data divide, and data changes the nature of knowledge)? How is privacy managed?

28. How will knowledge games not just measure and report on data but potentially shape our reality and ways of knowing?

29. What is lost or gained when we focus on using data-driven methodologies rather than other potential aspects of knowledge gaming, such as evoking emotion or incorporating ethnography? What is attained or forgotten when data is the only important text, rather than other aspects of humanity, such as poetry, sculpture, or discussion?

30. How do we ensure that data is not equated with knowledge?

Chapter 9. Knowledge

31. Do knowledge games force us to reevaluate how we create knowledge as a society but also our understanding of, and the purpose of, games? Are games still games if they produce knowledge?

32. How can our design and use of knowledge games influence how knowledge is created, who creates it, and how we can take action and make change based on it?

33. Who gets to know and make knowledge? Have we opened up the limits of knowledge production only temporarily because the limits of data have expanded—and so we need more people to participate? Will we close knowledge production once these limits are reached again? How is this reflected in our design of knowledge games?

34. Do knowledge games support engaged citizenship, or do they restrict it? How does our design influence this? Do we need to be active participants in knowledge games to be responsible citizens of the world?

35. How do we integrate values, ethics, and humanity into our design and use of knowledge games—both inside and outside games? How do we ensure that our fellow designers, players, educators, researchers, and even our computers and machines also integrate values into knowledge creation and use?

Introduction

1. The protocol, called restriction enzyme digestion, is a common procedure that has been done in biology laboratories for the past few decades. It involves using an enzyme to cut DNA at a specific site, called the restriction site. You can use this procedure to cut or cleave a bacteriophage and insert it in new DNA. You can also cleave DNA and put it into an electrophoresis gel so that the fragments separate. The procedure even helped lead to scientific discoveries such as recombinant DNA and the creation of new supplies of insulin for diabetics using *E. Coli* bacteria.

2. The special science program in my public high school afforded me this chance to apprentice in a science laboratory—an opportunity I have never had again, despite efforts to do so.

3. Galaxy Zoo (Zooniverse), www.galaxyzoo.org; eBird (National Audubon Society; Cornell Lab of Ornithology), http://ebird.org/content/ebird; Cicada Tracker (WNYC Radiolab), http://project.wnyc.org/cicadas (all accessed Feb. 23, 2015).

4. See, e.g., Nathan Prestopnik and Kevin Crowston, "Purposeful Gaming & Socio-Computational Systems: A Citizen Science Design Case," in *Group '12: Proceedings of the 17th ACM International Conference on Supporting Group Work* (New York: ACM, 2012). On the growth of citizen science projects, see Science for Citizens, SciStarter, http://scistarter.com (accessed Feb. 23, 2015), a database of over 100 citizen science projects. Under animal-related citizen science projects, for example, there are almost 100 projects listed. I also want to clarify that I am not dismissing the need for the rigorous training required to become a scientist, and the practice necessary to maintain this status, but noting that citizen scientists are also able to contribute to scientific knowledge through these projects.

5. Crowdsourcing is discussed in more detail in chapter 1.

6. Operation War Diary (Zooniverse; Imperial War Museum; The National Archives), www.operationwardiary.org; What's the Score at Bodleian? (Zooniverse; Bodleian Libraries, University of Oxford), www.whats-the-score.org; Kinsey Reporter (Kinsey Institute, Indiana University at Bloomington), www.kinseyreporter.org (all accessed Feb. 23, 2015).

7. *Foldit* (University of Washington Center for Game Science; UW Department of Biochemistry), http://fold.it/portal/ (accessed Feb. 23, 2015); *Reverse the Odds* (Stand up to Cancer; Cancer Research UK; Zooniverse; Maverick Television; Channel 4; Chunk), www .cancerresearchuk.org/support-us/citizen-science-apps-and-games-from-cancer-research-uk /reverse-the-odds (accessed July 2, 2015); *Monster Proof* (Verigames/DARPA), www .monsterproof.verigames.com/world (accessed Aug. 29, 2015); *Astro Drone Crowdsourcing Game* (Advanced Concepts Team, European Space Agency; Micro Air Vehicle Laboratory, TU Delft; Artificial Intelligence Group, Radboud University Nijmegen), www.esa.int/gsp /ACT/ai/projects/astrodrone.html (accessed Feb. 23, 2015). Augmented reality games such as the *Astro Drone Crowdsourcing Game* combine aspects of the virtual world or game content with the real world. Another example is *Ingress* (Google / Niantic Labs), www.ingress.com (accessed Aug. 1, 2015), a game that can be accessed from anywhere using an Android phone. People play on one of two teams and need to "capture" real locations around the world using their mobile device.

8. A. W. Combs, "Humanistic Education: Too Tender for a Tough World?" *Phi Delta Kappan* 62, no. 8 (Feb. 1981): 446–49.

9. See, e.g., David Shaffer, *How Computer Games Help Children Learn* (New York: Palgrave Macmillan, 2006).

10. "Video Game Industry Statistics [Infographic]," 2013, http://infographiclist.com/2013/02/20/video-game-industry-statistics-infographic-2; Entertainment Software Association, "2014 Essential Facts about the Computer and Video Game Industry," www.theesa.com/wp-content/uploads/2014/10/ESA_EF_2014.pdf; D. Takahashi, "Time Spent Playing Video Games Keeps Going Up," http://venturebeat.com/2010/03/02/time-spent-playing-video-games-keeps-going-up (all accessed Feb. 27, 2015); Jane McGonigal, *Reality Is Broken* (New York: Penguin, 2011). In a TED Conversation, McGonigal also observes that 5 million game players in the United States are spending more time per week playing games than a typical full-time job (40 hours) and that we are spending 3 billion hours a week globally playing games. See www.ted.com/conversations/44/we_spend_3_billion_hours_a_wee.html (accessed July 2, 2015).

11. Johnson, *Everything Bad Is Good for You* (New York: Riverhead Books, 2006), 17. Johnson quotes Dr. Spock, in an edition of his popular parenting book, writing of video games: "The best that can be said of them is that they may help promote eye-hand coordination in children. The worst that can be said is that they sanction, and even promote aggression and violent responses to conflict. But what can be said with much greater certainty is this: most computer games are a colossal waste of time" (17, quoting Benjamin Spock and Steven J. Parker, Dr. Spock's Baby and Child Care [New York: Pocket Books, 1998], 625). On the Gamergate controversy, see Kyle Wagner, "The Future of the Culture Wars Is Here, and It's Gamergate," *Deadspin*, Oct. 14, 2014, http://deadspin.com/the-future-of-the-culture-wars-is-here-and-its-gamerga-1646145844 (accessed Feb. 27, 2015). "Intimidation Game" (season 16, episode 14), originally aired on NBC's *Law and Order: SVU* on Feb. 11 2015.

12. On using games for social change and education, see Games for Change, www.gamesforchange.org, and the Learning, Education, and Games special interest group of the International Game Developers Association, www.igda.org/group/leg, as well as a series of edited collections written by the group, e.g., Karen Schrier, ed., *Learning, Education, and Games, Vol. 1: Curricular and Design Considerations* (Pittsburgh, PA: ETC Press, 2014). On examples of learning and games, see my introduction to Schrier, *Learning, Education, and Games, Vol. 1,* and games created by iCivics, BrainPOP, Filament Games, Learning Games Network, Making History, Schell Games, and Kognito. Research on gaming and learning includes James Gee, *What Games Have to Teach Us about Learning and Literacy* (New York: Palgrave Macmillan, 2003); Schrier, *Learning, Education, and Games, Vol. 1;* Shaffer, *How Computers Games Help Children Learn;* Kurt Squire, *Video Games and Learning: Teaching and Participatory Culture in the Digital Age* (New York: Teachers College Press, 2011); and Constance Steinkuehler, Kurt Squire, and Sasha Barab, eds., *Games, Learning, and Society: Learning and Meaning in the Digital Age* (New York: Cambridge University Press, 2012).

13. McGonigal, *Reality Is Broken.* See also her TED talk, "Gaming Can Make a Better World," Feb. 2010, www.ted.com/talks/jane_mcgonigal_gaming_can_make_a_better_world (accessed Sept. 3, 2015).

14. The terms "collective intelligence" and "distributed cognition" are discussed in more detail in chapter 5.

15. On use of the term "social participation games," see McGonigal, *Reality Is Broken.*

16. See, for instance, Katie Salen and Eric Zimmerman, *Rules of Play* (Cambridge, MA: MIT Press, 2003), who argue that all games can contribute to our knowledge of humanity in that we can always learn about humanity and our own world through the system that emerges through play, and the ways people behave and interact in that system. Thus, we

should let all games provide insight into humanity, for instance, as sites for research studies, through an evaluation of the emergent communities surrounding a game, or through reflection on one's play of a game. However, knowledge games seek to specifically help produce knowledge as part and parcel of the game's goals.

17. Note that as of August 2015, no researchers, developers, or designers label their games as knowledge games. The aim of this book is not to debate which name is the most accurate. Throughout these pages, I refer to the types of games that exist to encourage citizen participation in knowledge-making activities—like *Foldit*—as *knowledge games*. For more information on the naming debate, see a discussion on the Gamesnetwork listserv, Nov. 1, 2013, http://listserv.uta.fi/cgi-bin/wa?A0=GAMESNETWORK. Also, see more in chapter 1.

18. On games in education and curricular considerations, see Gee, *What Games Have to Teach Us*; Schrier, *Learning, Education, and Games, Vol. 1*; Shaffer, *How Computers Games Help Children Learn*; Squire, *Video Games and Learning*; and Steinkuehler, Squire, and Barab, *Games, Learning, and Society*. The term "serious games" refers to a movement that supports using games for so-called serious purposes, such as health issues, education, or government training. For more about this term, and debate surrounding it, see www.seriousgames .org. The term is intended to encapsulate all games that are meant not just for fun and leisure but for playing with a purpose, such as learning. See, e.g., P. Wouters, C. van Nimwegen, H. van Oostendorp, and E. D. van der Spek, "A Meta-analysis of the Cognitive and Motivational Effects of Serious Games," *Journal of Educational Psychology* 105, no. 2 (2013): 249.

19. See www.gamesforchange.org (accessed Feb. 24, 2015) and D. Ruggiero, "The Compass Rose of Social Impact Games," *International Journal of Computer and Electrical Engineering* 5, no. 6 (Dec. 2013): 597–601.

20. On revealing an argument, see, e.g., Ian Bogost's work on procedural rhetoric in *Unit Operations* (Cambridge, MA: MIT Press, 2008). In regard to games for change, note that I am not arguing that games for change are unable to spur real-world change, particularly as awareness is a first step in social change; rather, I am clarifying that what people typically refer to as "games for change" are not necessarily knowledge games. I also want to make clear that games for change can be very effective and are not necessarily a better or worse type of game than knowledge games; rather, they fulfill different needs.

Another related type of game is "engagement games" or "games for engagement," in which the actions in the game are actions with real-world referents. In other words, players are able to make real change and take action through the game. Eric Gordon, Stephen Walter, and Pablo Suarez describe engagement games as "a new type of interface for real-world processes . . . from community planning and data collection, disaster preparedness, advocacy and fundraising, to skill and network building." Gordon, Walter, and Suarez, "Engagement Games: A Case for Designing Games to Facilitate Real-World Action," http://engagementlab .emerson.edu/pdfs/engagement-game-guide.pdf, p. 7 (accessed May 7, 2015). For more about engagement games, also see the note in chapter 1.

21. Ian Bogost lists pranks, exercise, art, and work as main categories of functions of games in *How to Do Things with Videogames* (Minneapolis: University of Minnesota Press, 2011).

22. "Gamification" has no formal definition but has been applied to "everything from *Farmville* to LinkedIn's profile completeness bar" and involves adding "extrinsic reward— points, leaderboards, badges, and such . . . onto non-game experiences." John Ferrara, "Games for Persuasion: Argumentation, Procedurality, and the Lie of Gamification," *Games and Culture* 8, no. 4 (2013): 289–304, 291. Even though knowledge games and gamification are not synonymous, I acknowledge that there may be some overlap between the usages of the two concepts, particularly as definitions evolve popularly. Additionally, the boundaries between them may become more and more blurred as games in general evolve.

23. That said, there is some overlap between gamification and knowledge games, particularly because gamification has been applied to many different types of game and real-world experiences, and because the distinction between the "real world" and games becomes increasingly blurred. More about the overlap between these terms is considered in chapter 4.

24. On hierarchies of taste, see, e.g., Pierre Bourdieu, *Distinction: A Social Critique of the Judgement of Taste* (London: Routledge, 1984).

The phrase "not the message" is a play on Brenda Brathwaite and John Sharp's "The Mechanic Is the Message," in *Ethics and Game Design: Teaching Values through Play*, ed. Karen Schrier and David Gibson (Hershey, PA: IGI Global, 2009). This, in turn, is a play on Marshall McLuhan's famous phrase, "the medium is the message," from McLuhan, *Understanding Media: Extensions of Man* (1964; reprint, Cambridge, MA: MIT Press, 1994).

An ARG is a game experience that mixes fictional and nonfictional elements and typically involves multiple people. It can take "place in real time and evolve according to decisions taken by the players rather than by a programmer." Jeffrey Kim, Elan Lee, Timothy Thomas, and Caroline Dombrowski, "Storytelling in New Media: The Case of Alternate Reality Games, 2001–2009," *First Monday* 14, no. 6 (2009), http://firstmonday.org/ojs/index .php/fm/article/view/2484/2199 (accessed July 16, 2014). ARGs use a variety of tools to tell a story and provide puzzles to players, such as phone, video, websites, social media, and in-person events. The distributed audience often uses the Internet (such as through forums or social media) to collectively and collaboratively solve puzzles and progress through the story. ARGs first became popular in 2004 with *I Love Bees* (42 Entertainment), which was used to market *Halo 2*. Participants needed to solve puzzles and complete missions (using fake websites, phone calls, and the like) as part of a fictional story about a spaceship that crashed into Earth.

25. There are not many knowledge games that have been designed or used yet. Although I am critical in my analysis of these games throughout this book, we should also respect this growing field and realize the challenge of designing these types of experiences.

26. For an in-depth discussion of the distinction between disciplines and fields, see Sébastien Hock-Koon, conversation on GamesNetwork listserv, Feb. 9, 2015, http://listserv .uta.fi/cgi-bin/wa?A0=GAMESNETWORK.

27. While there are many other game-related considerations, such as aesthetic and technical concerns, identity and role playing, or story and narrative, I decided to focus on these three areas, as I argue that they are integral (though not exclusively so) to the effectiveness of knowledge games.

Chapter 1 · Contribution

1. Kinsey Reporter was created by the Kinsey Institute for Research in Sex, Gender, and Reproduction, and the Center for Complex Networks and Systems Research (CNetS) at Indiana University's iSchool, which was led by Filippo Menczer from CNetS and Julian Heiman from the Kinsey Institute, and originally released in 2012. Version 2.1.2 was released on Jan. 27, 2014. Most of the featured responses on the day I observed (Feb. 27, 2015) came from the United States and Europe, though it is open to participants in any country. The survey described was "Most Desired Sexual Activity for Valentine's Day," www.kinseyreporter.org/ explore# (accessed Feb. 27, 2015).

2. See eBird (Audubon Society; Cornell Lab of Ornithology), http://ebird.org/content /ebird/ (accessed Feb. 23, 2015), and ZomBee Watch, SFSU Dept. of Biology and Center for Computing for Life Sciences, and Natural History Museum of Los Angeles, www .zombeewatch.org (accessed Feb. 24, 2015). In relation to the Kinsey Report research, see "The Kinsey Reporter App: Reaching Out to the Mobile World," www.indiana.edu/~kinsey /newsletter/sp2013/kinseyreporternew.html (accessed Feb. 14, 2015).

3. Howe, "The Rise of Crowdsourcing," *Wired*, June 2006, http://archive.wired.com /wired/archive/14.06/crowds.html (accessed Feb. 24, 2015); Howe, *Crowdsourcing: Why the Power of the Crowd Is Driving the Future of Business* (New York: Crown, 2008), abstract at ACM Digital Library, http://dl.acm.org/citation.cfm?id=1481457 (accessed Sept. 3, 2015).

4. Daren Brabham, *Crowdsourcing* (Cambridge, MA, MIT Press, 2013), xix. For more on the platforms listed, see www.threadless.com, www.quora.com, and www.mturk.com (all accessed May 1, 2015).

5. Brabham, *Crowdsourcing*; "Facts & Stats," www.innocentive.com/about-innocentive /facts-stats (accessed Feb. 27, 2015).

6. Henry Jenkins, *Convergence Culture: Where Old and New Media Collide* (New York: NYU Press, 2006). "Spoiling" involves giving away clues as to a television show's ending or what happens in a show.

Approximately 3 million volunteers participated in the Tomnod campaign for the missing Malaysian Airlines flight MH370. Before the search was called off in the South China Sea and Gulf of Thailand areas, on Mar. 14, 2014, there were "more than 190 million map views, and each pixel available had been seen by a human eye at least 30 times" through Tomnod, according to Annalisa Merella, "Using Crowdsourcing to Search for Flight MH370 Has Both Pluses and Minuses," *Quartz.com*, Mar. 15, 2014, http://qz.com/188270/using-crowdsourcing-to -search-for-flight-mh-370-has-both-pluses-and-minuses (accessed Feb. 27, 2015). On MH370, see Michael Martinez and John Newsome, "Crowdsourcing Volunteers Comb Satellite Photos for Malaysia Airlines Jet," Mar. 12, 2014, www.cnn.com/2014/03/11/us/malaysia-airlines-plane -crowdsourcing-search, and "Missing Airplane: Malaysia Airlines MH370," www.tomnod .com/campaign/mh370_indian_ocean/map/1fsxcy1 (both accessed Feb. 27, 2015). Tomnod has also been used to look for Genghis Khan's tomb (see http://exploration.nationalgeographic .com/mongolia/expedition). In July 2015, debris suspected as being from the Malaysian jet MH370 washed ashore on some islands in the Indian Ocean, although its origins are still being debated.

On Tomnod's use in the earthquake in Nepal, see Laura Keeney, "DigitalGlobe Launches Map Crowdsourcing to Assist Nepal Relief," *Denver Post*, Apr. 27, 2015, www.denverpost.com /business/ci_27999415/digitalglobe-launches-map-crowdsourcing-assist-nepal-quake-relief; A. Marshall, "How Amateur Mappers Are Helping Recovery Efforts in Nepal," *CityLab*, Apr. 28, 2015, www.citylab.com/tech/2015/04/how-amateur-mappers-are-helping-recovery-efforts-in -nepal/391703/; Humanitarian Data Exchange (HDX) (UN Office for the Coordination of Humanitarian Affairs) https://data.hdx.rwlabs.org/; and Mark Wilson, "How the Candy Crush of Data Is Saving Lives in Nepal," *Fast Company*, Apr. 30, 2015,www.fastcodesign.com/3045699 /how-the-candy-crush-of-data-is-saving-lives-in-nepal (all accessed May 1, 2015).

7. See Ovia Fertility Tracker and Ovia Pregnancy Tracker (Ovuline), www.ovuline.com (accessed May 1, 2015).

8. Brabham, *Crowdsourcing*; www.dousaflavor.com (accessed Feb. 27, 2015).

9. The NYC MTA also recently released an app to help people share information on subway delays. For more about the Waze app, see www.waze.com/about (accessed July 2, 2015).

10. See, e.g., Mario Aguilar, "Police Are Trying to Undermine Waze with a Deluge of Phony Cop Sightings," *Gizmodo*, Feb. 12, 2015, http://gizmodo.com/police-are-trying-to -undermine-waze-with-a-deluge-of-ph-1685391655 (accessed Feb. 27, 2015).

11. Yuxiang Zhao and Qinghua Zhu, "Evaluation on Crowdsourcing Research: Current Status and Future Direction," *Information Systems Frontiers* 16, no. 3 (2012): 417–34. Zhao and Zhu explain that crowdsourcing is distinct from outsourcing and the crowd may not be involved so anonymously as the word suggests.

12. Nathan Prestopnik and Kevin Crowston, "Purposeful Gaming & Socio-Computational Systems: A Citizen Science Design Case," in *Group '12: Proceedings of the 17th ACM International*

Conference on Supporting Group Work (New York: ACM, 2012), 76; Prestopnik and Crowston, "Citizen Science System Assemblages: Understanding the Technologies That Support Crowdsourced Science," in *Proceedings of 2012 iConference, Toronto, Ontario, Canada, February 7–10, 2012* (New York: ACM, 2012); Sunyoung Kim, Jennifer Mankoff, and Eric Paulos, "Sensr: Evaluating a Flexible Framework for Authoring Mobile Data-Collection Tools for Citizen Science," in *Proceedings of the Annual ACM Computer Supported Cooperative Work Conference, San Antonio, Texas, February 23–27, 2013* (New York: ACM, 2013).

13. In *Happy Moths* (Citizen Sort), http://socs.ischool.syr.edu/happymatch/index.php /GameInitialization/startGame/59, players identify moths. In *Citizen Science*, developed by Kurt Squire from University of Wisconsin at Madison and Filament Games, funded by the National Science Foundation, players clean up lakes and learn about freshwater scientific issues (limnology). See www.gameslearningsociety.org/project-citizen-science.php. The *Gravestone Project* is part of EarthTrek, a program of the Geological Society of America. See www .geosociety.org/earthtrek/gravemap.aspx. (All accessed May 1, 2015).

14. Michael Mueller, Deborah Tippins, and Lynn Bryan, "The Future of Citizen Science," *Democracy and Education* 20, no. 1 (2012): 1–12; Prestopnik and Crowston, "Purposeful Gaming," 76. The term "citizen" is used because of its connection to the public or community, according to Mueller, Tippins, and Bryan: "Citizen science is incisively conceptualized as community-centered science, community service, participatory community-action research, street science, traditional ecological knowledge, social justice, scientific literacy, and humanistic science education" (12).

15. See "Project Overview," http://feederwatch.org/about/project-overview/ (accessed Feb. 24, 2015).

16. Mm^2 (NZ Marine Studies Centre, University of Otago), www.mm2.net.nz; Shark-Finder (JASON Learning), www.jason.org/citizen-science/sharkfinder; Mt. Diablo Morgan Fire (Nerds for Nature), http://nerdsfornature.org/monitor-change/diablo.html (all accessed Feb. 24, 2015).

17. LibCrowds (British Library), www.libcrowds.com, is built on top of PyBossa, an open-source framework for conducting citizen science and crowdsourcing projects; see PyBossa (Scifabric), http://pybossa.com and http://crowdcrafting.org/about. (All accessed Sept. 7, 2015.) SciStarter is a database of citizen science projects, put together by Science for Citizens LLC. For example, there are almost 100 animal-related citizen science projects listed. See http://scistarter.com. Citizen Science 2015 was held in San Jose, California. For more information, see http://citizenscienceassociation.org/conference/citizen-science-2015. (Both accessed Feb. 23, 2015).

18. Chiara Franzoni and Henry Sauermann, "Crowd Science: The Organization of Scientific Research in Open Collaborative Projects," *Research Policy* 43, no. 1 (2014): 1–20; Alexander J. Quinn and Benjamin B. Bederson, "Human Computation: A Survey and Taxonomy of a Growing Field," in *Proceedings of the Annual ACM CHI Conference, Vancouver, BC, Canada, May 7–21, 2011* (New York: ACM, 2011); Luis von Ahn, "Human Computation" (PhD diss., Carnegie Mellon University, 2005), 3, http://reports-archive.adm.cs.cmu.edu/anon /2005/cmu-cs-05-193.pdf (accessed May 1, 2015).

19. These examples of games were provided in Quinn and Bederson, "Human Computation." The *ESP Game* was developed by Luis von Ahn of Carnegie Mellon to help label images or provide metadata for images so computers could more quickly search through, find, and identify them. In the game, participants are randomly assigned to each other and cannot see what the other is typing. Each player needs to "guess" which tags ("strings") the other person is applying to an image. If they agree on the tags for the image, then each player gets points. There are also opportunities for bonus points. The "string" that the players typically agree on is also a good label for the image, according to Luis von Ahn and Laura Dab-

bish, "Labeling Images with a Computer Game," in *Proceedings of the Annual ACM CHI Conference, Vienna, Austria, April 24–29, 2004* (New York: ACM, 2004). Google bought and rebranded the game in 2011, changing its name to Google Image Labeler.

Luis von Ahn, Ben Maurer, Colin McMillen, David Abraham, and Manuel Blum originally created reCAPTCHA at Carnegie Mellon University. Its purpose was to both protect users from computer "spambots" trying to autofill a field or enter an area on a website and also to simultaneously digitize the words and numbers in books, street-view images, and other materials. If you sign up for a listserv or try to submit a comment on a website, you might get taken to a reCAPTCHA form where you first need to enter two sets of numbers or words on images to be able to gain access. Google acquired the service in 2009 (see more about the new reCAPTCHA at www.google.com/recaptcha/intro/index.html).

Soylent is an add-on to Microsoft Word in which anyone can invite and incorporate crowd contributions into their document to complete a variety of tasks, such as editing and proofreading or reviewing and fact-checking. See more at Michael Bernstein, Greg Little, Robert C. Miller, Björn Hartmann, Mark S. Ackerman, David Crowell, and Katrina Panovich, "Soylent: A Word Processor with a Crowd Inside," in *Proceedings of the Annual ACM UIST, New York, New York, October 3–6, 2010* (New York: ACM, 2010).

20. See the Games & Crowds conference agenda at https://docs.google.com/document/d /1AlJOkVsFF8PQVWUPMgBCDOBoumgqccXj3DCSRBMAAaE/edit (accessed July 2, 2015).

21. See, e.g., Brabham, *Crowdsourcing*.

22. See, e.g., Prestopnik and Kevin Crowston, "Purposeful Gaming," and von Ahn, "Human Computation." A benefit of the term "citizen science games" is that it implies knowledge production and builds on the long history of citizen involvement in contributing to scientific knowledge. The main downside of the term is that it is specifically focused on STEM knowledge production, which (1) attracts those interested in those topics but deters everyone else and (2) leaves out the possibility of contributing knowledge to other topics, such as humanistic, artistic, or social science pursuits. It's no surprise, as Brabham explains in *Crowdsourcing*, that a high percentage of people participating in science-related online crowdsourcing platforms such as InnoCentive are science-related professionals (93). This means that the term "citizen science" inadvertently excludes a proportion of the public who could be contributing valuable perspectives, information, and ways of seeing the world. These ideas are also discussed in Daren Brabham, "The Myth of Amateur Crowds: A Critical Discourse Analysis of Crowdsourcing Coverage," *Information, Communication & Society* 15, no. 3 (2012): 394–410, which also cites L. B. Jeppesen and K. R. Lakhani, "Marginality and Problem-Solving Effectiveness in Broadcast Search," *Organization Science* 21, no. 5 (2010): 1016–33, esp. 1026.

23. Eric Gordon, Stephen Walter, and Pablo Suarez, "Engagement Games: A Case for Designing Games to Facilitate Real-World Action," http://engagementlab.emerson.edu/pdfs /engagement-game-guide.pdf (accessed May 7, 2015), 7. Perhaps the knowledge games described in this book are a subtype of engagement games.

24. Jane McGonigal, *Reality Is Broken* (New York: Penguin, 2011).

25. David Shaffer. *How Computer Games Help Children Learn* (New York: Palgrave Macmillan, 2006), 94, italics his; David Shaffer, "Wag the Kennel: Games, Frames, and the Problem of Assessment," in *Handbook of Research on Effective Electronic Gaming in Education*, ed. Rick Ferdig (Hershey, PA: IGI Global, 2009).

26. Deborah Strumsky, Jose Lobo, and Joseph A. Tainter, "Complexity and the Productivity of Innovation," *Systems Research and Behavioral Science*, 27, no. 5 (2010): 496–509.

27. See more at E. Lofgren and Nina Fefferman, "The Untapped Potential of Virtual Game Worlds to Shed Light on Real World Epidemics," *Lancet Infectious Diseases* 7 (2007): 625–29, as well as Mark Ward, "Deadly Plague Hits World of Warcraft," BBC News, Sept. 5, 2005, http://news.bbc.co.uk/2/hi/technology/4272418.stm (accessed Aug. 18, 2015).

28. Brabham, *Crowdsourcing*, 43 (citing Howe, *Crowdsourcing*), 44–59.

29. SeeClickFix connects citizens with local governments, where the public can submit nonemergency issues, such as potholes or road hazards in their communities, and governments can manage the issues through the site. See http://en.seeclickfix.com/ (accessed May 1, 2015). Peer to Patent is an online platform whereby the public can participate in the review process and assess pending applications for patents. See New York Law School's overview at www.peertopatent.org (accessed Feb. 24, 2015).

30. Amazon's Mechanical Turk is an online exchange where people can find others to perform small tasks ("human intelligence tasks" or HITs) for them, such as transcribing interviews or responding to surveys. Most of these tasks are compensated at a very low rate. For more about Mechanical Turk, see www.mturk.com/mturk/welcome (accessed Feb. 24, 2015).

31. Andrea Wiggins and Kevin Crowston, "From Conservation to Crowdsourcing," in *Proceedings of the 45th Annual Hawaii International Conference on System Sciences, January 4–7, 2012, Maui, Hawaii* (Los Alamitos, CA: IEEE Computer Society, 2012), 2. Wiggins and Crowston define citizen science as "a form of research collaboration involving members of the public in scientific research projects to address real-world problems" (1).

32. See David Jonassen and Woei Hung, "All Problems Are Not Equal: Implications for Problem-Based Learning," *Interdisciplinary Journal of Problem-Based Learning* 2, no. 2 (2008): 6–28, and Woei Hung, "Team-Based Complex Problem Solving: A Collective Cognition Perspective," *Educational Technology Research and Development* 61 (2013): 365–84.

33. A MOBA is a multiplayer online battle area, which is genre that emerged from real-time strategy games. Games of this genre include *League of Legends* (Riot Games) and *Defense of the Ancients* (Valve Corporation).

34. *Dognition* may be another example of the cooperative contribution type, though it is arguably not a game but a crowdsourcing project. In *Dognition* (Canines Inc.), dog owners have the opportunity to play different games (such as one that tests whether your dog will follow her memory or your pointing). As the dog and owner play, there are questions the owner can answer through the online tool, regarding the dog's behavioral responses, for example. Pet owners then receive a detailed assessment of their dog's behavior and personality, while the *Dognition* creators get to gather data about dogs. See www.dognition.com/how-it -works (accessed Mar. 1, 2015).

35. In *Verb Corner* (Joshua Hartshorne / GameswithWords.org), participants need to complete tasks and answer questions related to linguistic meaning. For example, in the task "Fickle Folk," participants read a sentence and decide if the subject has changed his or her mind or beliefs based on the sentence structure (verbs and nouns are replaced with fake words). Players earn points for each completed task and end up on a very basic leaderboard. See www.gameswithwords.org/VerbCorner (accessed May 1, 2015).

In *Ignore That!* (Joshua Hartshorne / GameswithWords.org), participants first decide which side of an arrow a plus sign is on and then whether a word is in an orange or white font (testing the Stroop effect). The purpose is to understand whether people can attend to information despite distractions. At the end of the experiment, you learn how many you got "incorrect" and your timing on the question, but there is no way to advance in the game, and the game does not change, so it is more like a quiz or online research experiment rather than a game. *Ignore That!* tests people's ability to deal with the Stroop effect (e.g., judging that a word is colored orange even if it says white), so there are some "right" responses (such as identifying the correct color despite distractions). However, the game is testing the individual interpretations or abilities of people, so it could be categorized as analysis distribution rather than in the cooperative contribution category. See www.gameswithwords.org/IgnoreThat (accessed May 1, 2015).

For more on *Who Is the Most Famous?* and *Apetopia*, see introduction, chapter 2, and chapter 5.

36. *FASTT Math* (Houghton Mifflin Harcourt) supports math fluency and facility with math facts (such as the multiplication table). The game adapts to the player based on the player's abilities and progress in the game. See www.hmhco.comproducts/fastt-math/. *Nevermind* (Flying Mollusk) uses the Oculus Rift virtual reality head-mounted apparatus and measures a player's biological feedback (such as heart rate) to adapt the game to the player. If a person's heart rate increases, the game gets harder and scarier, so the player needs to learn how to stay calm to be successful. See http://store.steampowered.com/app/342260/. (Both accessed May 1, 2015.)

Chapter 2 · Design

1. *Reverse the Odds* (Stand up to Cancer, Cancer Research UK, Zooniverse, Maverick Television, Channel 4, and Chunk), www.cancerresearchuk.org/support-us/citizen-science -apps-and-games-from-cancer-research-uk/reverse-the-odds (accessed July 2, 2015). There are more than 30 different citizen science projects on the Zooniverse platform, including Galaxy Zoo; Condor Watch, www.condorwatch.org, where volunteers can look at images to identify California condors and annotate their location, their social behavior, and their distance to food sources; or Notes from Nature, www.notesfromnature.org, where volunteers help transcribe museum records on minerals, plants, fungi, and animal specimens and data.

2. *Foldit* (Center for Game Science at University of Washington and UW Department of Biochemistry), http://fold.it/portal/ (accessed Feb. 23, 2015).

3. Michael A. Nielsen, *Reinventing Discovery: The New Era of Networked Science* (Princeton, NJ: Princeton University Press, 2012); F. Khatib et al., "Algorithm Discovery by Protein Folding Game," *Proceedings of the National Academy of Sciences* 108, no. 47 (Nov. 22, 2011): 18949–53. There are more than 20 different types of amino acids, which are the building blocks of protein. While this would amount to many unique combinations, only an estimated 50,000 to 100,000 proteins are absolutely essential to the functioning of the human body, D. Davison of Bristol-Myers Squibb estimates the number as 84,000 proteins. See Daniel B. Davison, "The Number of Human Genes and Proteins," www.nsti.org/publications /MSM/2002/pdf/330.pdf (accessed Aug. 18, 2015). See also "Gaming for a Cure," Aug. 4, 2010, www.washington.edu/news/2010/08/04/gaming-for-a-cure-computer-gamers-tackle -protein-folding/ (accessed Nov. 18, 2013).

4. Adrien Treuille, quoted in "Why Video Games Are Key to Modern Science," *CNN .com*, Oct. 23, 2011, www.cnn.com/2011/10/23/tech/innovation/foldit-game-science-poptech /index.html (accessed Oct. 15, 2013). Greg Costikyan contends that games are distinct from toys in "I Have No Words and I Must Design," www.costik.com/nowords.html (accessed Nov. 18, 2013). But I argue that when Treuille uses "toy-like" in this context, he implies that the proteins are beyond being simply a toy but that they become playful and meaningful objects to manipulate because of the game context.

5. With the phrase "constructivist activities," I allude to Seymour Papert's "constructivism" ("children as builders of their own intellectual structures"), which builds on Piaget's notion that kids can actively construct knowledge. *Mindstorms: Children, Computers, and Powerful Ideas*, 2nd ed. (New York: Basic Books, 1993), 7, citing Jean Piaget. For an example of Piaget's work, see M. Boden, *Piaget* (London: Harvester Press, 1979). Papert explains that kids can build knowledge through actual building, particularly collaboratively and in scaffolded, situated experiences.

The term "communities of practice" refers to a body of work from the late 1980s and early 1990s, particularly research by Jean Lave and Etienne Wenger, which considers the social and communal aspects of learning. Lave and Wenger argue that learning is situated in gatherings of people around common interests, passions, or goals, including everything from students in schools to engineering societies. See Jean Lave and Etienne Wenger, *Situated Learning: Legitimate Peripheral Participation* (Cambridge: Cambridge University Press, 1991), and Wenger, "Introduction to Communities of Practice," http://wenger-trayner.com /introduction-to-communities-of-practice/. Others have specifically considered the situated nature of learning, and how experiential learning contributes to cognition, including J. S. Brown, A. Collins, and P. Duguid, "Situated Cognition and the Culture of Learning," *Educational Researcher* 18, no. 1 (Jan.–Feb. 1989): 32–42. More about communities of practice is discussed in chapters 5 and 7.

6. Nielsen, *Reinventing Discovery*, 146.

7. For more about this idea of epistemic frames, see David W. Shaffer's discussion of epistemic games in *How Computer Games Help Children Learn* (New York: Palgrave, 2006). On the Mason-Pfizer monkey virus discovery, see F. Khatibet et al., "Crystal Structure of a Monomeric Retroviral Protease Solved by Protein Folding Game Players," *Nature Structural & Molecular Biology* 18, no. 10 (Oct. 2011): 1175–77.

8. Doug Church, "Formal Abstract Design Tools," in *The Game Design Reader: A Rules of Play Anthology*, ed. Katie Salen and Eric Zimmerman (Cambridge, MA: MIT Press, 2005); Kurt Squire, *Video Games and Learning: Teaching and Participatory Culture in the Digital Age* (New York: Teachers College Press), 2011. The problem space is what David Jonassen, "Toward a Design Theory of Problem Solving," *Educational Technology Research and Development* 48, no. 4 (2000): 63–84, calls the "mental representation of the situation in the world." Ibid., 65, citing A. Newell and H. Simon, *Human Problem Solving* (Englewood Cliffs, NJ: Prentice Hall, 1972). More about the connection between problem space and games is discussed in chapter 3.

9. On games as dynamic systems, see Katie Salen and Eric Zimmerman, *Rules of Play* (Cambridge, MA: MIT Press, 2003); Shaffer, *How Computer Games Help Children Learn*.

10. See Jonathan Mahler, "The White and Gold (No, Blue and Black!) Dress That Melted the Internet," *New York Times*, Feb. 27, 2015, www.nytimes.com/2015/02/28/business/a-simple -question-about-a-dress-and-the-world-weighs-in.html, and Adam Rogers, "The Science of Why No One Agrees on the Color of This Dress," *Wired.com*, Feb. 26, 2015, www.wired.com /2015/02/science-one-agrees-color-dress/ (accessed July 2, 2015).

11. *Specimen* (PepRally, LLC), https://itunes.apple.com/us/app/specimen-a-game-about -color/id999930535; *Apetopia* (Kai-Uwe Barthel), http://colors.htw-berlin.de (accessed May 1, 2015).

12. Kai-Uwe Barthel, Nico Hezel, Moritz Klack, Andy Lindemann, Christopher Moller, and Christine Wiederer, "Perceived Color Difference," www.ifaf-berlin.de/fileadmin/docs /Projekte/VIRPA/Barthel_Perceived_Color_Difference.pdf (accessed Aug. 1, 2015).

13. Some people may think I am being a little too fast and loose with my interdisciplinary definition of design. For more nuanced, specific definitions, see Dan Saffer, *Designing for Interaction: Creating Innovative Applications and Devices*, 2nd ed. (Berkeley, CA: New Riders, 2009), and Salen and Zimmerman, *Rules of Play*.

14. See Eric Zimmerman, "Play as Research: The Iterative Design," http://ericzimmerman .com/files/texts/Iterative_Design.htm (accessed May 1, 2015).

15. Playtesting involves having players play your game, watching them play the game, and inviting their feedback. Playtesting can become iterative testing if you continue to test, assess, and revise your game's design. See also Zimmerman, "Play as Research."

16. For an explanation of design principles, see Saffer, *Designing for Interaction*.

17. J. Huizinga, "Nature and Significance of Play as a Cultural Phenomenon," and R. Caillois, "The Definition of Play: The Classification of Games," quoted in Jesper Juul, *Half-*

Real: Video Games between Real Rules and Fictional Worlds (Cambridge, MA: MIT Press, 2011), 13.

18. Juul, *Half-Real*, 1, 36.

Chapter 3 · Problem Solving

1. On *EteRNA* (Carnegie Mellon University; Stanford University), see http://eterna.cmu .edu/web/about/ (accessed Feb. 24, 2015). The RNA nucleotides (building blocks of RNA protein) are A, C, G, and U, or adenine, cytosine guanine, and uracil. Through the tutorial missions, the player learns that there are different pairings between these nucleotides, like G with C, A with U, and G with U. The G to C bond is the strongest.

2. In *EteRNA*, there are a lot of rules provided in a short amount of time in the tutorial, and it can be overwhelming. It's a lot to keep track of for a layperson—even for a molecular biology lover and former laboratory researcher like myself (20 years prior). Learning these rules might have been facilitated through leveling, such as providing only one new rule every round or every few rounds. This is one of the particular benefits of games—helping to make complex content more digestible. Players in *Dragon Age: Inquisition* (BioWare), for instance, can learn how to create complicated alchemy (creation of potions), or investigate weapons upgrades because the interface scaffolds this complexity, and the player embraces each new feature of the game one by one. We get a lot of complex rules about the RNA molecules but no context or information as to why, so it's hard to figure out the best way to approach each problem, except through trial and error. Each challenge has a number of goals associated with it, which you need to reach to successfully submit your design to the system. As you work, you receive feedback, such as "great pairings," to know you are on the right track. You can also receive hints along the way. The more difficult challenges, however, can be extremely hard to solve, and it's not always clear how much you need to practice before you take on those harder puzzles. Often there are many different ways to solve a challenge, particularly as you move to harder missions.

3. Jeehyung Lee et al., "RNA Design Rules from a Massive Open Laboratory," *Proceedings of the National Academy of Sciences* 111, no. 6 (2014): 2122–27. On October 3, 2014, at 1:02 p.m. there were 44 players logged in on *EteRNA*. For more about EteRNABot, see http:// eternabot.org.

4. On *The SUDAN Game* (USC GamePipe Lab; Carnegie Mellon), see Peter Landwehr, Marc Spraragen, Balki Ranganathan, Kathleen M. Carley, and Michael Zyda, "Games, Social Simulations, and Data—Integration for Policy Decisions: The SUDAN Game," *Simulation and Gaming* 44, no. 1 (2013): 151–77, www.cs.cmu.edu/~plandweh/pdfs/GandS_SUDANgame .pdf (accessed May 1, 2015). As of September 2015, I am not aware of *The SUDAN Game* being released for public play, and I have not been able to play it.

5. David Jonassen and Woei Hung, "All Problems Are Not Equal: Implications for Problem-Based Learning," *Interdisciplinary Journal of Problem-Based Learning*, 2, no. 2 (2008): 6–28.

6. David Jonassen, "Toward a Design Theory of Problem Solving," *Educational Technology Research and Development*, 48, no. 4 (2000): 63–84, 65.

7. Ibid.

8. Ibid., citing Newell and Simon, *Human Problem Solving*.

9. Doug Church, "Formal Abstract Design Tools," in *The Game Design Reader: A Rules of Play Anthology*, ed. Katie Salen and Eric Zimmerman (Cambridge, MA: MIT Press, 2005); Kurt Squire, *Video Games and Learning: Teaching and Participatory Culture in the Digital Age* (New York: Teachers College Press, 2011).

10. Raph Koster, *A Theory of Fun for Game Design* (Sebastopol, CA: O'Reilly, 2004). See also Squire, *Video Games and Learning*, 4, 8.

11. Squire, *Video Games and Learning*, 94.

12. Ian Bogost, "Reality Is Alright: A Review of Jane McGonigal's Book *Reality Is Broken*," http://bogost.com/blog/reality_is_broken/ (accessed May 5, 2015).

13. More formally, puzzles are not the same as problems. Puzzles typically have one solution, whereas a problem could have many different possible solutions, according to Squire, *Video Games and Learning*, 90–91. In regard to games with problems that have no clear solutions, I argue that if a problem continues to be unsolved, the problem space does not cease to be, and it does not necessarily suggest the problem (or the game) was ill designed.

14. Jonassen, "Toward a Design Theory of Problem Solving."

15. Ibid., 73. Andrew Walker and Heather Leary expand on this idea: "Logical problems are highly constrained and generally abstracted, such as drawing four straight lines that intersect all points in a 3x3 array of dots. They rely on a single approach to reasoning to unlock the solution (in this particular case, a willingness to draw a triangle that goes outside the bounds of the dot array, which is then bisected by the fourth line)." Walker and Leary, "A Problem-Based Learning Meta Analysis: Differences across Problem Types, Implementation Types, Disciplines, and Assessment Levels," *Interdisciplinary Journal of Problem-Based Learning*, 3, no. 1 (2009): 6–28, 16.

16. *Phylo* (McGill University; Jérôme Waldispühl), http://phylo.cs.mcgill.ca/#!/EN/About (accessed July 2, 2015).

17. "Algorithmic problems cover the formulas you might expect for symbol manipulation domains like math or physics but also things like recipes for cooking. They tend to focus on following appropriate steps to arrive at a solution state." Walker and Leary, "Problem-Based Learning Meta Analysis," 16.

18. Jonassen, "Toward a Design Theory of Problem Solving," 77.

19. Allen Newell and Herbert A. Simon, *Human Problem Solving* (Englewood Cliffs, NJ: Prentice Hall, 1972), as cited in Rodney L. Custer, "Examining the Dimensions of Technology," *International Journal of Technology and Design Education* 5, no. 5 (1995): 219–44.

20. On ill-structured problems, see, e.g., H. S. Barrows, "Is It Truly Possible to Have Such a Thing as dPBL?" *Distance Education* 23, no. 1 (2002): 119–22, and Walker and Leary, "Problem-Based Learning Meta Analysis." On multiple solutions, see Cindy E. Hmelo-Silver and H. S. Barrows, "Goals and Strategies of a Problem-Based Learning Facilitator," *Interdisciplinary Journal of Problem-Based Learning* 1, no. 1 (2006): 21–39.

21. Walker and Leary, "Problem-Based Learning Meta Analysis." As described in the chapter, dilemmas could involve open questions such as how to ethically and responsibly handle a student who cheats.

22. J. Introne, R. Laubacher, G. Olson, and T. Malone, "Solving Wicked Social Problems with Socio-computational Systems," *Künstliche Intelligenz* 27, no. 1 (2013): 45–52, quote at 45; Horst Rittel and M. Webber, "Dilemmas in a General Theory of Planning," *Policy Science* 4 (1973): 155–69; Detlef Schoder, Johannes Putzke, Panagiotis Takis Metaxas, Peter A. Gloor, and Kai Fischbach, "Information Systems for 'Wicked Problems'—Research at the Intersection of Social Media and Collective Intelligence," *Business & Information Systems Engineering* 6, no. 1 (2014): 3. Examples of potential wicked problems are from Jeffrey B. Liebman, "Building on Recent Advances in Evidence-Based Policymaking" (Results for America and The Hamilton Project at the Brookings Institute, 2013), www.brookings.edu/~/media/research/files/papers/2013/04/17-liebman-evidence-based-policy/thp_liebmanf2_413.pdf (accessed May 1, 2015). For more about wicked problems, see www.wickedproblems.com/1_wicked_problems.php.

23. Minhong Wang, Bian Wu, Kinshuk, Nian-Shing Chen, and J. Michael Spector, "Connecting Problem-Solving and Knowledge-Construction Processes in a Visualization-Based Learning Environment," *Computers & Education* 68 (Oct. 2013): 293–306.

24. See, e.g., Koster, *Theory of Fun for Game Design*, and Squire, *Video Games and Learning.*

25. James Gee, *What Games Have to Teach Us about Learning and Literacy* (New York: Palgrave Macmillan, 2003). Also see Luis von Ahn, "Human Computation" (PhD diss., Carnegie Mellon University, 2005).

26. "The underlying complexity of wicked problems comes from the fact that they are problems complicated by social interactions that are fluid, evolving, and involve conflicting interests. Resolving wicked problems requires parallel discourse, multiple iterations, changes of beliefs, and unpredictable revisions. Outcomes may be emergent and depend on the intensity, quality, and perception of contributions over time and may never be final or 'true' in an absolute, agreed-upon sense." Schoder et al., "Information Systems for 'Wicked Problems,'" 4. Also see Luis von Ahn, "Human Computation" (PhD diss., Carnegie Mellon University, 2005).

27. Luis von Ahn calls human computation "a paradigm for utilizing human processing power to solve problems that computers cannot yet solve." Von Ahn, "Human Computation," 3. According to Schoder et al., "Its power comes from its use of games designed to engage humans in collaborating, sometimes without their knowledge, . . . human beings acting as sensors, thus exercising what is called human sensing, crowd data sensing or public data sensing . . . a wave of (human) social networks and structures are emerging as important drivers for the development of novel communication and computing paradigms." Schoder et al., "Information Systems for 'Wicked Problems,'" 6.

28. David Shaffer, *How Computer Games Help Children Learn* (New York: Palgrave Macmillan, 2006), 65; Karen Schrier, "Ethical Thinking and Sustainability in Role-Play Participants: A Preliminary Study," *Simulation & Gaming*, published online before print, Dec. 17, 2014, http://sag.sagepub.com/content/early/2014/11/20/1046878114556145.refs (accessed Aug. 19, 2015).

29. "Supporting deliberation, that is, the 'process where communities (1) identify possible solutions for a problem, and (2) select the solution(s) from this space that best meet(s) their diverse needs' (Klein 2012, p. 449); helping knowledge workers navigate social graphs (link prediction, identifying relevant individuals, assessing the strength of ties, assessing the embeddedness and position of individuals, etc.); highlighting relevant documents based on their processing through social interactions (who is using or working on documents, in which social position and context, and does this indicate relevance?); exploiting human computation; creating individualized information cockpits that monitor topical domains in a customized way (including hot topic identification and predictive capabilities of how things, items, issues, people, etc. may evolve); and coping with large collections of semistructured or unstructured data, technically as well as semantically." Schoder et al., "Information Systems for 'Wicked Problems,'" 5, quoting Mark Klein, "Enabling Large-Scale Deliberation Using Attention-Mediation Metrics," *Computer Supported Cooperative Work* 21 (2012): 449–73.

30. Woei Hung, "Team-Based Complex Problem Solving: A Collective Cognition Perspective," *Educational Technology Research and Development* 61 (2013): 365–84.

31. Yigal Rosen, "Comparability of Conflict Opportunities in Human-to-Human and Human-to-Agent Online Collaborative Problem Solving," *Technology, Knowledge, and Learning* 19 (2014): 147–64.

32. Ibid.

33. Karim R. Lakhani, Lars Bo Jeppesen, Peter A. Lohse, and Jill A. Panetta, "The Value of Openness in Scientific Problem Solving," working paper, 2007, www.hbs.edu/faculty/publication%20files/07-050.pdf (accessed Aug. 1, 2015).

34. Ibid.

35. Deniz Eseryel, Xun Ge, Dirk Ifenthaler, and Victor Law, "Dynamic Modeling as a Cognitive Regulation Scaffold for Developing Complex Problem-Solving Skills in an Educational Massively Multiplayer Online Game Environment," *Journal of Educational Computing Research* 45, no. 3 (2011): 265–86, which specifically used as models of self-determination theory and self-efficacy theory to consider how the students represented a problem, motivational factors, and the task itself.

36. Richard M. Ryan, C. Scott Rigby, and Andrew Przybylski, "The Motivational Pull of Video Games: A Self-Determination Theory Approach," *Motivation and Emotion* 30 (2006): 347–63. Regarding research time related to practicing a problem, see John R. Anderson, *Cognitive Psychology and Its Implications*, 6th ed. (New York: Worth, 2005).

37. S. Buckingham Shum et al. explain: "Constructing a narrative around a factual conclusion elevates raw information to the level of insight and connects it to the archetypal human experience . . . Narratives can contextualise information in order to highlight the human implications of factual data, connecting it to our shared reality through conflict, character, and plot. Ultimately, storytelling allows us to transform unapproachable scientific data and factual conclusions into the common language that has been used for communication throughout the evolution of the human species." Shum et al., "Towards a Global Participatory Platform," *European Physical Journal Special Topics* 214 (2012): 109–52, 135. Regarding a story context, see David H. Jonassen, "Designing Research-Based Instruction for Story Problems," *Educational Psychology Review* 15, no. 3 (2003): 267–96.

38. "Monster Proof: A Game Design for Verification," July 29, 2015, www.verigames.com /news/2015/07/29/monster-proof-game-design-for-verification (accessed Aug. 20, 2015). In their newer *Monster Proof*, the designers decided to "embrace the math" and instead include mathematical Suduko-like puzzles in the game. For *Stormbound* and *Monster Proof* (TopCoder, Inc. / Verigames / DARPA), see www.verigames.com.

39. *Fraxinus* (Sainsbury Laboratory; Genome Analysis Centre; John Innes Centre), https://apps.facebook.com/fraxinusgame (accessed July 2, 2015), is similar to *Phylo* in that you are working with gene sequences, but the metaphor it uses is trees and leaves rather than blocks. The goal of *Fraxinus* is to compare genetic data that is represented by colored patterns to provide insight into how Chalara fungus is infecting ash trees. *Fraxinus* is very challenging, and because it is a Facebook game, you can also see which of your friends have played it and where they score on a friends leaderboard.

40. Shum et al. offer three specific scenarios for the use of story in problem-solving platforms, like knowledge games, including one based on narrative. "Narrative search results: the generation of exploratory interfaces and search results which connect heterogeneous elements meaningfully into a narrative, using semantic templates and natural language generation; the use of narratological models to underpin story-based annotation and browsing: the derivation of a story markup scheme in the design of a prototype 'storybase' for healthcare knowledge sharing[;] and, the distinctive role of narrative as a form of knowledge representation for complex systems thinking: complex systems make sense in retrospect, as analysts seek to construct plausible narratives for each other and decision-makers to make sense of complex systems. Narrative has an important place in some of the most influential work on sensemaking support systems." Shum et al., "Towards a Global Participatory Platform," 135.

Forgotten Island (Syracuse University Students / Nathan Prestopnik; Encyclopedia of Life; DiscoverLife.org; National Science Foundation), www.citizensort.org/web.php/games /startforgottenisland (accessed May 2, 2015), is a point-and-click adventure game that motivates participants to classify specimens of animals and plants, and complete tasks for a robot. The game uses narrative and storytelling techniques (the game begins, "On a mysterious island, somewhere in the middle of an unknown ocean . . ." and an online graphic novel-type introduction); however, it is not always clear how the story is integrated with the act of classifying specimens.

41. For more about brain areas, see Anderson, *Cognitive Psychology and Its Implications*, 254–55; K. Christoff, V. Prabhakaran, J. Dorfman, Z. Zhao, J. K. Kroger, K. J. Holyoak, and J. D. Gabrieli, "Rostrolateral Prefrontal Cortex Involvement in Relational Integration during Reasoning," *Neuroimage* 14 (2001): 1136–49. For more about context and meaning and how to integrate them, see Shum et al., "Towards a Global Participatory Platform."

42. "People are fixed on representing an object according to its conventional function and fail to represent its novel function." Anderson, *Cognitive Psychology and its Implication*, 270. Anderson provides the example of a tack box that people are focused on using as a box rather than a stand for a candle, suggesting that people are biased by experiences to "prefer certain operators when solving a problem." Ibid., 271, citing K. Duncker, "On Problem-Solving," trans. L. S. Lees, *Psychological Monographs* 58, no. 270 (1945).

43. See, e.g., Gee, *What Games Have to Teach Us*.

44. See, e.g., Shaffer, *How Computer Games Help Children Learn*.

45. Gee, *What Games Have to Teach Us*, 90; on using games to test hypotheses, see, e.g., Hercules Panoutsopoulos and Demetrios G. Sampson, "A Study on Exploiting Commercial Digital Games into School Context," *Educational Technology & Society* 15, no. 1 (2012): 15–27.

46. J. R. Savery and T. M. Duffy, "Problem Based Learning: An Instructional Model and Its Constructivist Framework," *Educational Technology* 35 (1995): 31–38.

47. Shaffer, *How Computer Games Help Children Learn*. Shaffer describes a simulation as a real or virtual representation of part of the world. Squire explains that, technically, models are different from simulations in that they are static rather than dynamic, but he uses the terms interchangeably. Squire, *Video Games and Learning*, 23. Likewise, in addition to serving as models of the world or simulations of problems, Gee explains how games situate meanings and relationships and problems in "embodied experiences to solve problems and reflect on the social relationships and identities in the modern world." Gee, *What Games Have to Teach Us*, 48.

48. Seymour Papert, *Mindstorms: Children, Computers, and Powerful Ideas*, 2nd ed. (New York: Basic Books, 1993). For more about microworlds, see Papert, *Mindstorms*, and "Micro-world," Edutech Wiki, http://edutechwiki.unige.ch/en/Microworld (accessed Aug. 1, 2015).

49. See previous note on models and simulations. Also, Squire, in *Video Games and Learning*, maintains that games and simulations are not the same. Simulations create a dynamic interplay of the variables of a particular process. Games also provide goals, roles, experimentation with identity, agency, character progression, competition, story, choice, and consequence. Games are both media and play, so they need to be seen in relation to books and TV, as well as other games, such as board games, toys, and sports. For the general properties of games, see Squire, *Video Games and Learning*, and Shaffer, *How Computer Games Help Children Learn*.

50. Even mental practice and simulation can be effective. David H. Jonassen explains that mental practice and simulation is useful because "episodic future thought is dependent on recollection of the past experiences . . . They are episodic in nature . . . For everyday situations, event cues (e.g., restaurant, party) cause us to mentally generate hypothetical future scenarios. For example, what are the most likely results of my engaging that person at a party in some conversation? The ability to simulate the future is related to one's ability to remember the past. Mental simulations enhance open-mindedness, making it easier to generate and evaluate alternatives." Jonassen, "Designing for Decision Making," *Education Technology Research and Development* 60 (2012): 354.

51. Karen Schrier, "Emotion, Empathy, and Ethical Thinking in *Fable III*," in *Emotions, Technology, and Digital Games*, ed. Sharon Tettegah and Wen-Hao Huang (New York: Elsevier, 2015); "Ethical Thinking and Sustainability."

52. See, for instance, Shaffer, *How Computer Games Help Children Learn*; Squire, *Video Games and Learning*; and Daniel Aronson, "Overview of Systems Thinking," www.thinking .net/Systems_Thinking/OverviewSTarticle.pdf (accessed Aug. 20, 2015).

53. Shaffer, *How Computer Games Help Children Learn*, 5. Shaffer paraphrases what John Dewey might say by explaining that "these tasks were not about life, they were life itself" (6), citing John Dewey, *School and Society* (Chicago: University of Chicago Press, 1915).

54. Shaffer explains that "the knowledge that matters is the knowledge that experts have," so that what is valued to know is both socially constructed and determined by a game's design. Ibid., 68.

55. See, e.g., Shaffer, *How Computer Games Help Children Learn*, and Gee, *What Games Have to Teach Us*.

56. For how knowledge is stored in objects, characters, and environments, see Gee, *What Games Have to Teach Us*, 109–11. About breaking up larger problems into smaller ones, Anderson explains that you can create sub-goals to "eliminate the difference between the current state and the condition for applying a desired operator." *Cognitive Psychology and Its Implications*, 261. See also Newell and Simon, *Human Problem Solving*.

57. Anderson, *Cognitive Psychology and Its Implications*; Lakhani et al., "Value of Openness in Scientific Problem Solving."

58. Deniz Eseryel, Dirk Ifenthaler, and Xun Ge, "Towards Innovation in Complex Problem Solving Research: An Introduction to the Special Issue," *Education Technology Research and Development* 61 (2013): 359–63; John D. Bransford, Ann L. Brown, and Rodney R. Cocking, *How People Learn: Brain, Mind, Experience, and School* (Washington, DC: National Academy Press, 1999).

59. Squire, *Video Games and Learning*, discusses how games and designers shape the world and the biases related to this.

60. Ibid.

Chapter 4 · Motivation

1. On *Eyewire* (Seung Lab), see http://blog.eyewire.org/about/ (accessed Feb. 25, 2015). Formerly the Seung Lab was housed at MIT, where it was launched in 2012. As of July 2015, it is at Princeton University.

2. As you click around and color the brain cell in *Eyewire*, the cell pieces turn green or red (in the game), depending on whether you colored within the lines properly or not. A progress bar also lets you know how much of the neuron cell you have traced. Once you feel you have colored in the cell properly, you can click complete and get an accuracy percentage and feedback on which parts of the cell you missed (shown in yellow in the game).

3. See http://blog.eyewire.org/about/. The development team holds weekly competitions in addition to the *EyeWire* game playing you can do at your leisure.

4. James Gorman, "Recruiting Help: Gamers," *New York Times*, May 26, 2014, www .nytimes.com/2014/05/27/science/recruiting-help-gamers.html (accessed Feb. 25, 2015); Jinseop S. Kim et al., "Space-Time Wiring Specificity Supports Direction Selectivity in the Retina," *Nature* 509 (2014): 331–36.

5. Deniz Eseryel, Victor Law, Dirk Ifenthaler, Xun Ge, and Raymond Miller, "An Investigation of the Interrelationships between Motivation, Engagement, and Complex Problem Solving in Game-based Learning," *Educational Technology & Society* 17, no. 1 (2014): 42–53; Selen Turkay, Daniel Hoffman, Charles K. Kinzer, Pantiphar Chantes, and Christopher Vicari, "Toward Understanding the Potential of Games for Learning: Learning Theory, Game

Design Characteristics, and Situating Video Games in Classrooms," *Computers in the Schools: Interdisciplinary Journal of Practice, Theory, and Applied Research* 31 (2014): 2–22; Katie Salen and Eric Zimmerman, *Rules of Play* (Cambridge, MA: MIT Press, 2003).

6. According to Zynga's 2014 annual investor report, "We process and serve more than 300 terabytes of content for our players every day. We continually analyze game data to optimize our games. We believe that combining data analytics with creative game design enables us to create a superior player experience." 2014 Annual Report (Form 10-K), http:// investor.zynga.com/secfiling.cfm?filingID=1193125-14-62902 (accessed May 2, 2015), 3.

7. *Grand Theft Auto V* sold approximately 26.75 million units in 2013. "The 10 Bestselling Video Games of 2013," www.thefiscaltimes.com/Media/Slideshow/2013/12/13/10-Bestselling -Video-Games-2013; www.vgchartz.com (both accessed May 2, 2015).

8. Richard M. Ryan and Edward L. Deci, "Intrinsic and Extrinsic Motivations: Classic Definitions and New Directions," *Contemporary Educational Psychology* 25 (2000): 54–67; M. Jordan Raddick et al., "Galaxy Zoo: Motivations of Citizen Scientists," *Astronomy Education Review* 9 (2010): 2.

9. Ryan and Deci, "Intrinsic and Extrinsic Motivations," quotes at 55.

10. The rate is 25.3% for the year ending in September 2014 (the Bureau of Labor Statistics collects data from September 2013 to September 2014). "Volunteering in the United States, 2014," www.bls.gov/news.release/volun.nr0.htm (accessed Mar. 1, 2015).

11. E. Gil Clary and Mark Snyder, "The Motivations to Volunteer: Theoretical and Practical Considerations," Current Directions in Psychological Science 8 (1999): 156–59, 156; E. Gil Clary, Mark Snyder, Robert D. Ridge, John Copeland, Arthur A. Stukas, Julie Haugen, and Peter Miene, "Understanding and Assessing the Motivations of Volunteers: A Functional Approach," *Journal of Personality and Social Psychology* 74, no. 6 (1998): 1516–30.

12. Samuel Shye, "The Motivation to Volunteer: A Systemic Quality of Life Theory," *Social Indicators Research* 98 (2010): 183–200.

13. Ibid., quote at 198.

14. John Wilson, "Volunteerism Research: A Review Essay," *Nonprofit and Voluntary Sector Quarterly* 41, no. 2 (2012): 176–212, quote at 190.

15. Ibid.

16. Nancy Morrow-Howell, Song-Iee Hong, and Fengyan Tang, "Who Benefits from Volunteering? Variations in Perceived Benefits," *Gerontologist* 49, no. 1 (2009): 91–102.

17. Dana Rotman et al., "Dynamic Changes in Motivation in Collaborative Citizen-Science Projects," in *Proceedings of the Annual ACM Computer Supported Cooperative Work Conference, Seattle, Washington, February 11–15, 2012* (New York: ACM, 2012); C. Daniel Batson and Nadia Ahmad, "Four Motives for Community Involvement," *Journal of Social Issues* 58, no. 3 (2002): 217–26.

18. Lakhani et al. explain that one-third of the problems posed in their research were able to be solved through the InnoCentive mechanism (the site boasts an overall 85% completion rate). Karim R. Lakhani, Lars Bo Jeppesen, Peter A. Lohse, and Jill A. Panetta, "The Value of Openness in Scientific Problem Solving," working paper, 2007, www.hbs.edu/faculty /publication%20files/07-050.pdf (accessed Aug. 1, 2015); "Facts & Stats," www.innocentive .com/about-innocentive/facts-stats (accessed Aug. 1, 2015). The average reward for the solutions studied in this research was almost $30,000, and problems posited through the platform typically have a time limit.

19. Lakhani et al., "Value of Openness in Scientific Problem Solving."

20. Daren Brabham, "Moving the Crowd at iStockphoto: The Composition of the Crowd and Motivations for Participation in a Crowdsourcing Application," *First Monday* 13, no. 6 (2008), http://firstmonday.org/article/view/2159/1969 (accessed Feb. 25, 2015).

21. Daren Brabham, "Moving the Crowd at Threadless," *Information, Communication, & Society* 13, no. 8 (2010): 1122–45.

22. On other motivations uncovered, see Rotman et al., "Dynamic Changes in Motivation"; on contradictory findings and context, see Daren Brabham, *Crowdsourcing* (Cambridge, MA: MIT Press, 2013).

23. Yuxiang Zhao and Qinghua Zhu, "Evaluation on Crowdsourcing Research: Current Status and Future Direction," *Information Systems Frontiers* 16 (2014): 417–34.

24. Rotman et al., "Dynamic Changes in Motivation," quote at 217.

25. Ibid., quote at 217. The website www.biotrackers.net was no longer accessible as of Feb. 25, 2015.

26. Rotman et al., "Dynamic Changes in Motivation."

27. Anne Bowser, Derek Hansen, Yurong He, Carol Boston, Matthew Reid, Logan Gunnell, and Jennifer Preece, "Using Gamification to Inspire New Citizen Science Volunteers," in *Gamification 2013: Proceedings of the First International Conference on Gameful Design, Research, and Applications: October 2–4, 2013, Stratford, Ontario, Canada* (New York: ACM, 2013); Rotman et al., "Dynamic Changes in Motivation."

28. Raddick et al., "Galaxy Zoo: Motivations of Citizen Scientists."

29. Ioanna Iacovides, Charlene Jennett, Cassandra Cornish-Trestrail, and Anna L. Cox, "Do Games Attract or Sustain Engagement in Citizen Science? A Study of Volunteer Motivations," in *CHI 2013: Changing Perspectives: Conference Proceedings of the 31st Annual CHI Conference on Human Factors in Computing Systems: 27 April–2 May 2013, Paris, France* (New York: ACM, 2013), 1105.

30. Ibid.

31. Nathan Prestopnik and Kevin Crowston, "Purposeful Gaming & Socio-Computational Systems: A Citizen Science Design Case," in *Group '12: Proceedings of the 17th ACM International Conference on Supporting Group Work* (New York: ACM, 2012).

32. Ibid.

33. On *Forgotten Island*, see chapter 3. See also Nathan Prestopnik and Dania Souid, "Forgotten Island: A Story-Driven Citizen Science Adventure," in *CHI 2013: Changing Perspectives: Conference Proceedings of the 31st Annual CHI Conference on Human Factors in Computing Systems: 27 April–2 May 2013, Paris, France* (New York: ACM, 2013).

34. Rotman et al., "Dynamic Changes in Motivation."

35. Turkay et al., "Toward Understanding the Potential of Games for Learning," referencing R. Stevens, T. Satwicz, and L. McGarthy, "In Game, in Room, in World: Reconnecting Video Game Play to the Rest of Kids' Lives," in *The Ecology of Games*, ed. Katie Salen (Cambridge, MA: MIT Press, 2008), 41–46.

36. Brian Burke, *Gamify* (Brookline, MA: Bibliomotion, 2014); Jane McGonigal, *Reality Is Broken* (New York, NY: Penguin, 2011), 37; Paul Darvasi, "Escaping the Asylum: The Art of Transforming Your Class into a Living Video Game," presented at Games in Education Conference, Troy, NY, Aug. 6, 2014. An ARG is a game experience that mixes fictional and nonfictional elements and typically involves multiple people and take "place in real time and evolve according to decisions taken by the players rather than by a programmer." Jeffrey Kim, Elan Lee, Timothy Thomas, and Caroline Dombrowski, "Storytelling in New Media: The Case of Alternate Reality Games, 2001–2009," *First Monday* 14, no. 6 (2009), http://firstmonday.org/ojs/index.php/fm/article/view/2484/2199 (accessed July 16, 2014).

37. Richard M. Ryan, C. Scott Rigby, and Andrew Przybylski, "The Motivational Pull of Video Games: A Self-Determination Theory Approach," *Motivation and Emotion* 30 (2006): 348, citing Richard Bartle, "Hearts, Clubs, Diamonds, Spades: Players Who Suit MUDs," 1996, http://mud.co.uk/richard/hcds.htm (accessed May 2, 2015), and *Designing Virtual Worlds* (Indianapolis, IN: New Riders, 2004).

38. Nick Yee, *The Proteus Paradox* (New Haven, CT: Yale University Press, 2014).

39. Marino and Hayes contend that "research in inclusive middle school science class-rooms in the United States indicates that students' motivation to play the games is directly linked to the game context and their personal characteristics. For example, boys consistently rate games where they take command of a virtual environment (e.g., piloting a ship) higher than girls. On the other hand, girls appear to prefer games that involve social interaction more than boys do. However, both boys and girls rate games that address world health issues equally. Clearly additional research in this area is needed to examine how student level char-acteristics, including gender, reading ability, and race across cultures, respond to different types of gameplay narratives." Matthew T. Marino and Michael T. Hayes, "Promoting Inclu-sive Education, Civic Scientific Literacy, and Global Citizenship with Videogames," *Cultural Studies of Science Education* 7 (2012): 952.

40. "Crowley and Jacobs (2002) suggest that islands of expertise based on understanding of and interest in a specific topic create 'abstract and general themes' that students are able to use in other contexts. Here I add that islands of expertise include development of identity and adoption of practices associated with the ways of knowing of a particular community. That is, I argue that islands of expertise are organized around coherent epistemic frames, and that these frames—these ways of looking at the world associated with different communities of practice— are the 'abstract and general themes' that students use to leverage experience in an island of ex-pertise in new situations." David Shaffer, *How Computer Games Help Children Learn* (New York: Palgrave Macmillan, 2006), 232, citing Kevin Crowley and M. Jacobs, "Building Islands of Expertise in Everyday Family Activity," in *Learning Conversations in Museums*, ed. G. Lein-hardt, K. Crowley, and K. Knutson (Mahwah, NJ: Lawrence Erlbaum, 2002).

41. Thomas W. Malone, "Toward a Theory of Intrinsically Motivating Instruction," *Cog-nitive Science* 4 (1981): 333–69; Daniel Pink, *Drive: The Surprising Truth about What Moti-vates Us* (New York: Riverhead Books, 2011).

42. Ryan, Rigby, and Przybylski, "Motivational Pull of Video Games"; Yee, *Proteus Para-dox*; Mihaly Csikszentmihalyi, *Beyond Boredom and Anxiety: The Experience of Play in Work and Games* (San Francisco: Jossey-Bass, 1975), xiii; McGonigal, *Reality Is Broken*, 36.

43. Turkay et al., "Toward Understanding the Potential of Games for Learning," refer-encing L. A. Leotti, S. S. Iyengar, and K. N. Ochsner, "Born to Choose: Biological Bases for the Need for Control," *Trends in Cognitive Science* 14 (2010): 457–63; Salen and Zimmerman, *Rules of Play*.

44. For the idea that choices matter, see Yee, *Proteus Paradox*, and Turkay et al., "Toward Understanding the Potential of Games for Learning"; for the connection between choices and learning, see D. I. Cordova and M. R. Lepper, "Intrinsic Motivation and the Process of Learning: Beneficial Effects of Contextualization, Personalization, and Choice," *Journal of Educational Psychology* 19 (1996): 715–30.

45. Becta, "Computer Games in Education Project: Report," 2002, http://tna.europarchive .org/20080509164701/partners.becta.org.uk/index.php?section=rh&rid=13595, and "Computer Games in Education Project: Aspects," 2002, http://tna.europarchive.org/20080509164701 /http://partners.becta.org.uk/index.php?section=rh&&catcode=&rid=13588, 2001 (both ac-cessed May 2, 2015); Malone, "Toward a Theory of Intrinsically Motivating Instruction"; Turkay et al., "Toward Understanding the Potential of Games for Learning."

46. Deniz Eseryel, Victor Law, Dirk Ifenthaler, Xun Ge, and Raymond Miller, "An Inves-tigation of the Interrelationships between Motivation, Engagement, and Complex Problem Solving in Game-Based Learning," *Educational Technology & Society* 17, no. 1 (2014): 42–53.

47. Amanda Lenhart, Joseph Kahne, Ellen Middaugh Sr., Alexandra Rankin Macgill, Chris Evans, and Jessica Vitak, "Teens, Video Games, and Civics" (Pew Internet & American Life Project, Sept. 16, 2008), www.pewinternet.org/files/old-media/Files/Reports/2008/PIP

_Teens_Games_and_Civics_Report_FINAL.pdf.pdf (accessed May 2, 2015); Nick Yee, "Motivations for Play in Online Games," *CyberPsychology and Behavior* 9, no. 6 (2006): 772–75; Yee, *Proteus Paradox.*

48. Malone, "Toward a Theory of Intrinsically Motivating Instruction," and Thomas Malone, "What Makes Things Fun to Learn? A Study of Intrinsically Motivating Computer Games," Aug. 1980, http://cci.mit.edu/malone/tm%20study%20144.pdf (accessed July 2, 2015).

49. Nathan Prestopnik and Dania Souid, "Forgotten Island."

50. Malone, "Toward a Theory of Intrinsically Motivating Instruction."

Chapter 5 · Social Interaction

1. *Who Is the Most Famous?* (Yoni Alter, Cory Forsyth, and Rob O'Rourke), http://whoisthemostfamous.com (accessed Feb. 25, 2015).

2. When I played, responses to "Who is the most famous Kelly?" were that 55% said Kelly Clarkson, 7% said Kelly Osbourne, 5% said Kelly Rowland, and only 4% said Kelly Ripa, indicating that the demographic is more likely 13- to 25-year-olds, rather than my demographic, 35-to-50-year-old women. Based on a play-through of *Who Is the Most Famous?* on Oct. 20, 2014.

3. Based on a play-through of *Who Is the Most Famous?* on Oct. 20, 2014.

4. Constance Steinkuehler and Sean Duncan, "Scientific Habits of Mind in Virtual Worlds," *Journal of Science Education and Technology* 17, no. 530 (2008): 530–43; C. A. Anderson, D. A. Gentile, and K. E. Dill, "Prosocial, Antisocial, and Other Effects of Recreational Video Games," in *Handbook of Children and the Media*, 2nd ed., ed. D. G. Singer and J. L. Singer (Thousand Oaks, CA: Sage, 2012); David Shaffer, *How Computer Games Help Children Learn* (New York: Palgrave Macmillan, 2006); Kurt Squire, *Video Games and Learning: Teaching and Participatory Culture in the Digital Age* (New York: Teachers College Press, 2011). In relation to humans and computers working together, when are human beings, alongside computers or other people, particularly powerful and effective, and when are they less effective? Is our goal to build artificially intelligent machines or to find ways to more intimately connect computers and human beings to each other? See, e.g., Walter Isaacson, *The Innovators* (New York: Simon & Schuster, 2014).

5. See Constance Steinkuehler, "The Game of Life; Online Gaming Is Far from the Anti-Intellectual, Anti-Social Time Waster It Is Commonly Thought to Be," *Ottawa Citizen*, Sept. 3, 2007, and Anderson, Gentile, and Dill, "Prosocial, Antisocial, and Other Effects of Recreational Video Games." "Make Love, Not Warcraft," originally aired on Comedy Central's *South Park* on Oct. 4, 2006; see http://southpark.cc.com/full-episodes/s10e08-make-love-not-warcraft (accessed May 2, 2015).

6. Barbara M. Hall, "Designing Collaborative Activities to Promote Understanding and Problem-Solving," *International Journal of e-Collaboration* 10, no. 2 (2014): 5–71; Noriko Hara, Paul Solomon, Seung-Lye Kim, and Diane H. Sonnenwald, "An Emerging View of Scientific Collaboration," *Journal of the American Society for Information Science and Technology* 54 (2003): 952–65, quote at 953, citing Michael Schrage, *No More Teams!: Mastering the Dynamics of Creative Collaboration* (New York: Currency Doubleday, 1995), 33; G. Stahl, T. Koschmann, and D. Suthers, "Computer-Supported Collaborative Learning: An Historical Perspective," in *The Cambridge Handbook of the Learning Sciences*, ed. R. K. Sawyer (Cambridge: Cambridge University Press, 2006)

7. See more in Stahl, Koschmann, and Suthers, "Computer-Supported Collaborative Learning," 411. See also P. Dillenbourg, "What Do You Mean by 'Collaborative Learning'?" in *Collaborative Learning: Cognitive and Computational Approaches*, ed. P. Dillenbourg, 1–16 (Amsterdam: Elsevier Science; New York: Pergamon, 1999).

8. "Collaboration is a process by which individuals *negotiate and share meanings* relevant to the problem-solving task at hand . . . collaboration is a coordinated, synchronous activity that is the result of a continued attempt to construct and maintain a shared conception of a problem." Stahl, Koschmann, and Suthers, "Computer-Supported Collaborative Learning," 411 (emphasis added by authors) citing J. Roschelle and S. Teasley, "The Construction of Shared Knowledge in Collaborative Problem Solving," in *Computer Supported Collaborative Learning*, ed. C. E. O'Malley (Heidelberg: Springer-Verlag, 1995), 70.

9. Hall, "Designing Collaborative Activities," citing K. Mäkitalo-Siegl, *Interaction in Online Learning Environments: How to Support Collaborative Activities in Higher Education Settings* (Saarbrücken: Lambert, 2009), 9.

10. Pierre Lévy, *Collective Intelligence: Mankind's Emerging World in Cyberspace* (Cambridge, MA: Perseus, 1997); E. F. Fama, "Efficient Capital Markets: A Review of Theory and Empirical Work," *Journal of Finance* 25, no. 2 (1970): 383–417; Terry Flew and Sal Humphreys, "Games: Technology, Industry, Culture," in *New Media: An Introduction*, 2nd ed., ed. Terry Flew (South Melbourne: Oxford University Press, 2005), 101–14; John A. L. Banks, "Negotiating Participatory Culture in the New Media Environment: Auran and the Trainz Online Community—An (Im)possible Relation," Melbourne DAC 2003, http://hypertext.rmit.edu.au/dac/papers/Banks.pdf (accessed May 6, 2015); Henry Jenkins, *Convergence Culture: Where Old and New Media Collide* (New York: NYU Press, 2006). The title of this section is taken from a song from an episode of the PBS show *Daniel Tiger's Neighborhood*.

11. On the lack of a common definition, see Detlef Schoder, Johannes Putzke, Panagiotis Takis Metaxas, Peter A. Gloor, and Kai Fischbach, "Information Systems for 'Wicked Problems'—Research at the Intersection of Social Media and Collective Intelligence," *Business & Information Systems Engineering* 6, no. 1 (2014): 5; Don Tapscott and Anthony Williams, *Wikinomics: How Mass Collaboration Changes Everything* (New York: Portfolio, 2006); "About the MIT Center for Collective Intelligence," http://cci.mit.edu/about_mitcenter.html (accessed July 2, 2015).

12. Celia B. Harris, Amanda J. Barnier, John Sutton, and Paul G. Keil, "Couples as Socially Distributed Cognitive Systems: Remembering in Everyday Social and Material Contexts," *Memory Studies* 7, no. 3 (2014): 285–97, 286–87.

13. Woei Hung, "Team-Based Complex Problem Solving: A Collective Cognition Perspective," *Educational Technology Research and Development* 61 (2013): 365–84, citing E. Salas and S. M. Fiore, "Why Team Cognition: An Overview," in *Team Cognition: Understanding the Factors That Drive Process and Performance*, ed. E. Salas and S. M. Fiore (Washington, DC: American Psychology Association, 2004), and S. W. J. Kozlowski and D. R. Ilgen, "Enhancing the Effectiveness of Work Groups and Teams," *Psychological Science in the Public Interest* 7 (2006): 77–124, 81. "By studying the navigation team on a navy ship . . . and an air plane pilot crew in the cockpit . . . , Hutchins illustrated how team cognition is divided by the functions (e.g. navigation plotting or navigation bearing time) required for navigating a ship through a water and systematically distributed to and performed by the respective crew members (e.g. navigation plotter or navigation bearing recorder/timer). Then, collectively, the integration of the performance of navigation plotter, navigation bearing recorder/timer, captain, and other crew members results in a success or failure of navigating through water." Hung, "Team-Based Complex Problem Solving," 370, citing Edwin Hutchins, "How a Cockpit Remembers Its Speed," *Cognitive Science* 19 (1995): 265–88.

14. James Hollan, Edwin Hutchins, and David Kirsh, "Distributed Cognition: Toward a New Foundation for Human-Computer Interaction Research," *ACM Transactions on Computer-Human Interaction* 7, no. 2 (2000): 174–96; Hutchins, "How a Cockpit Remembers Its Speed"; Graham Button, "Against Distributed Cognition," *Theory Culture, & Society* 25,

no. 2 (2008): 87–104; James Gee, *What Games Have to Teach Us About Learning and Literacy* (New York: Palgrave Macmillan, 2003), 184–85.

15. Gee, *What Games Have to Teach Us*; also see chapters 1 and 3.

16. James Surowiecki, *The Wisdom of Crowds* (New York: Random House, 2005). Surowiecki argues, however, that experts can have biases and blind spots that render them unable to always make the best judgment. Also, it's really hard to always know who the true expert is.

17. Gee, *What Games Have to Teach Us*; Stefan Wuchty, Benjamin F. Jones, and Brian Uzzi, "The Increasing Dominance of Teams in Production of Knowledge," *Science* 316 (2007): 1036–39; Karim R. Lakhani, Lars Bo Jeppesen, Peter A. Lohse, and Jill A. Panetta, "The Value of Openness in Scientific Problem Solving," working paper, 2007, www.hbs.edu/faculty/publication%20files/07-050.pdf (accessed Aug. 1, 2015).

18. On social constructivism, see Lev S. Vygotsky, *Mind in Society: The Development of Higher Psychological Processes* (Cambridge, MA: Harvard University Press, 1978); on social learning theory, see Albert Bandura, *Social Learning Theory* (Englewood Cliffs, NJ: Prentice Hall, 1977); on situated learning theory, see Jean Lave and Etienne Wenger, *Situated Learning: Legitimate Peripheral Participation* (New York: Cambridge University Press, 1991).

19. Selen Turkay, Daniel Hoffman, Charles K. Kinzer, Pantiphar Chantes, and Christopher Vicari, "Toward Understanding the Potential of Games for Learning: Learning Theory, Game Design Characteristics, and Situating Video Games in Classrooms," *Computers in the Schools: Interdisciplinary Journal of Practice, Theory, and Applied Research* 31 (2014): 2–22; Bandura, *Social Learning Theory*; On *Way* specifically, see K. Schrier and D. Shaenfield, "Collaboration and Emotion in *Way*," in in *Emotions, Technology, and Digital Games*, ed. Sharon Tettegah and Wen-Hao Huang (New York: Elsevier, 2015); *Way* (Coco & Co.), www.makeourway.com (accessed May 6, 2015).

20. On communities of practice, see Lave and Wenger, *Situated Learning*. According to Zagal and Bruckman, "A community of practice involves a collection of individuals sharing mutually defined practices, beliefs, and understandings over an extended time frame in the pursuit of a shared enterprise." Jose P. Zagal and Amy Bruckman, "Designing Online Environments for Expert/Novice Collaboration: Wikis to Support Legitimate Peripheral Participation," *Convergence* 16, no. 4 (2010): 451–70, 452. James Gee and Elisabeth Hayes describe affinity spaces as places where "people affiliate with others based primarily on shared activities, interests, and goals." Gee and Hayes, "Nurturing Affinity Spaces and Game-Based Learning," in *Games, Learning, and Society: Learning and Meaning in the Digital Age*, ed. Constance Steinkuehler, Kurt Squire, and Sasha Barab (New York: Cambridge University Press, 2012), 67.

21. See Constance Steinkuehler, "Introduction to Section II: Games as Emergent Culture," in Steinkuehler, Squire, and Barab, *Games, Learning, and Society*, 123–28; Katie Salen and Eric Zimmerman, *Rules of Play* (Cambridge, MA: MIT Press, 2003); and Katie Salen and Eric Zimmerman, eds., *The Game Design Reader: A Rules of Play Anthology* (Cambridge, MA: MIT Press), 2005.

22. Salen and Zimmerman, *Game Design Reader*, 41. Kurt Squire discusses how players form ideological worlds with cultural values, ethics, norms that are negotiated and interpreted through the community in "Open-Ended Video Games: A Model for Developing Learning for the Interactive Age," in *The Ecology of Games: Connecting Youth, Games, and Learning*, ed. Katie Salen (Cambridge, MA: MIT Press), as cited in Turkay et al., "Toward Understanding the Potential of Games for Learning," and Matthew J. Gaydos and Kurt D. Squire, "Role Playing Games for Scientific Citizenship," *Cultural Studies of Science Education* 7 (2012): 821–44. See also Gee, *What Games Have to Teach Us about Learning and Literacy*.

23. Gee and Hayes, "Nurturing Affinity Spaces and Game-Based Learning"; Salen and Zimmerman, *Rules of Play*; C. A. Steinkuehler, "Massively Multiplayer Online Gaming as a Constellation of Literacy Practices," *E-Learning* 4, no. 3 (2007): 297–318, 301. "World of Warcraft players socialize within the game, and they come together in fan sites to discuss, critique, analyze, and mod the game." Gee and Hayes, "Nurturing Affinity Spaces and Game-Based Learning," 130. See also Rebecca W. Black and Stephanie M. Reich, "A Sociocultural Approach to Exploring Virtual Worlds," https://webfiles.uci.edu/rwblack/VirtualLitsProofs.pdf (accessed May 9, 2015).

24. Kurt Squire, "From Content to Context: Video Games as Designed Experiences," *Educational Researcher* 35, no. 8 (2006): 19–29, as cited in Turkay et al., "Toward Understanding the Potential of Games for Learning"; Squire, *Video Games and Learning*, 12.

25. See, e.g., Hans van der Meij, Eefje Albers, and Henny Leemkuil, "Learning from Games: Does Collaboration Help," *British Journal of Educational Journal* 42, no. 4 (2011): 655–64; Deniz Eseryel, Victor Law, Dirk Ifenthaler, Xun Ge, and Raymond Miller, "An Investigation of the Interrelationships between Motivation, Engagement, and Complex Problem Solving in Game-based Learning," *Educational Technology & Society* 17, no. 1 (2014): 42–53; Nick Yee, *The Proteus Paradox* (New Haven, CT: Yale University Press, 2014).

26. P. Resnick, "Beyond Bowling Together: Sociotechnical Capital," in *Human-Computer Interaction in the New Millennium*, ed. J. M. Carroll (New York: ACM; Boston: Addison-Wesley, 2002); Jane McGonigal, *Reality Is Broken* (New York: Penguin, 2011).

27. Van der Meij, Albers, and Leemkuil, "Learning from Games."

28. Ikram Bououd and Imed Boughzala, "A Serious Game Supporting Team Collaboration," Proceedings of Conference on Computer Supported Cooperative Work, San Antonio, Texas, February 23–27, 2013, http://ieeexplore.ieee.org/xpl/articleDetails.jsp?arnumber=6285843 (accessed July 2, 2015); Ikram Bououd and Imed Boughzala, "The Design of a Collaboration-Oriented Serious Game," presented at the 2012 International Conference on Communications and Information Technology (ICCIT), June 26–28, 2012; Alice Mitchell and Carol Savill-Smith, "The Use of Computer and Video Games for Learning: A Review of the Literature" (Learning and Skills Development Agency, 2004), www.m-learning.org/docs/The%20use%20of%20 computer%20and%20video%20games%20for%20learning.pdf (accessed July 2, 2015); K. Inkpen, "Gender Differences in an Electronic Games Environment," in *Proceedings of the ED-MEDIA 94 World Conference on Educational Multimedia and Hypermedia, Vancouver, Canada, 25–30 June 1994* (Charlottesville, VA: Association of Advancement of Computing in Education, 1994).

29. María-Esther Del-Moral and Alba-Patricia Guzmán-Duque, "*CityVille*: Collaborative Game Play, Communication, and Skill Development in Social Networks," *New Approaches in Educational Research* 3, no. 1 (2014): 11–19.

30. Paul Darvasi, "Escaping the Asylum: The Art of Transforming Your Class into a Living Video Game," presented at Games in Education Conference, Troy, NY, Aug. 6, 2014.

31. Hayeon Song, Jihyun Kim, Kelly E. Tenzek, and Kwan Min Lee, "The Effects of Competition and Competitiveness upon Intrinsic Motivation in Exergames," *Computers in Education* 29, no. 4 (2013): 1702–8; Lena Pareto, Magnus Haake, Paulina Lindström, Bjorn Sjödén, and Agneta Gulz, "A Teachable-Agent-Based Game Affording Collaboration and Competition: Evaluating Math Comprehension and Motivation," *Education Technology Research and Development* 60 (2012): 723–51.

32. Lave and Wenger, *Situated Learning*; John Seely Brown, Allan Collins, and Paul Duguid, "Situated Cognition and the Culture of Learning," *Educational Researcher* 18 (1989): 32–42; Minhong Wang, Bian Wu, Kinshuk, Nian-Shing Chen, and J. Michael Spector, "Connecting Problem-Solving and Knowledge-Construction Processes in a Visualization-Based Learning Environment," *Computers & Education* 68 (2013): 293–306, quote at 294.

33. Karen Schrier, "Designing Digital Games to Teach History," in *Learning, Education, and Games, Vol. 1: Curricular and Design Considerations*, ed. K. Schrier (Pittsburgh, PA: ETC Press, 2014); "Using Augmented Reality Games to Teach 21st Century Skills," in *Proceedings of ACM SIGGRAPH 2006 Educators Program, Boston, MA* (New York: ACM, 2006); and "Reliving History with *Reliving the Revolution*: Using Augmented Reality Games To Teach," in *Games and Simulations in Online Learning: Research and Development Frameworks*, ed. M. Prensky, C. Aldrich, and D. Gibson (Hershey, PA: IGI, 2006). Depending on which role they inhabited (such as a female loyalist or Minuteman soldier), participants would receive slightly different testimonials from the virtual historic characters in the game. For example, Paul Revere would divulge more information to a fellow Minuteman soldier than to a female loyalist. This factored into how the participants would judge and evaluate the evidence they received.

34. K. J. Topping, *The Peer Tutoring Handbook: Promoting Co-operative Learning* (London: Croom Helm; Cambridge, MA: Brookline, 1988); Turkay et al., "Toward Understanding the Potential of Games for Learning"; Constance Steinkuehler and Yoonsin Oh, "Apprenticeship in Massively Multiplayer Online Games," in Steinkuehler, Squire, and Barab, *Games, Learning, and Society*.

35. See, e.g., work on cognitive tutors in John R. Anderson, A. T. Corbett, Ken Koedinger, and R. Pelletier, "Cognitive Tutors: Lessons Learned," *Journal of Learning Sciences* 4 (1995): 167–207, and X. Bai, John B. Black, and J. Vitale, "Learning with the Assistance of a Reflective Agent," in *Agent-Based Systems for Human Learning and Entertainment*, ed. R. Axtell, G. Fagiolo, S. Kurihara, H. Nakashima and A. Namatame (New York: Association for Computing Machinery); Lave and Wenger, *Situated Learning*; Brown, Collins, and Duguid, "Situated Cognition and the Culture of Learning"; Wang et al., "Connecting Problem-Solving and Knowledge-Construction Processes"; and Thomas Malaby, "Parlaying Value: Capital in and beyond Virtual Worlds," *Games and Culture* 1 (2006): 141.

36. Steinkuehler and Oh, "Apprenticeship in Massively Multiplayer Online Games," define an apprenticeship relationship that happens through a community of practice, whereas there is a master who teaches and an apprentice who learns.

37. According to Gee, the jigsaw method involves one individual or team having expertise on one part of a topic, content, or gameplay, and working with other people who have complementary expertise. Gee, *What Games Have to Teach Us*, 191.

38. G. Stahl and F. Hesse, "Paradigms of Shared Knowledge," *International Journal of Computer-Supported Collaborative Learning* 4, no. 4 (2009): 365–69.

39. Hung continues, "The scale and the complexity of most real life problems today are beyond one person's physical, cognitive, and sometimes professional expertise capacities." Hung, "Team-Based Complex Problem Solving," 267.

40. Mark Klein, "Enabling Large-Scale Deliberation Using Attention-Mediation Metrics," *Computer Supported Cooperative Work* 21 (2012): 449–73; Surowiecki, *Wisdom of Crowds*.

41. As of May 8, 2015, there were 23,457 ideas submitted and 549 ideas implemented on Dell's IdeaStorm; see www.ideastorm.com. On the tragedy of the commons, see R. De Young, "Tragedy of the Commons," in *Encyclopedia of Environmental Science*, ed. D. E. Alexander and R. W. Fairbridge (Hingham, MA: Kluwer Academic, 1999), www.personal.umich.edu/~rdeyoung/tragedy.html#sthash.8v6ka6yN.dpuf (accessed July 2, 2015). On the "long tail," see Chris Anderson, www.longtail.com/about.html (accessed May 10, 2015).

42. Klein, "Enabling Large-Scale Deliberation." "Unlike a debate, students do not have to take sides and persuade others, but are free to explore positions flexibly and to make concessions." E. Michael Nussbaum, "Collaborative Discourse, Argumentation, and Learning: Preface and Literature Review," *Contemporary Educational Psychology* 33 (2008): 345–59, 349.

43. Nussbaum, "Collaborative Discourse, Argumentation, and Learning," 347. On dialogic processes, see D. Hicks, "Contextual Inquiries: A Discourse-Oriented Study of Classroom Learning," in *Discourse, Learning, and Schooling*, ed. D. Hicks (New York: Cambridge University Press, 1996), and D. Kuhn, W. Goh, K. Iordanou, and D. Shaenfield, "Arguing on the Computer: A Microgenetic Study of Developing Argument Skills in a Computer-Supported Environment," *Child Development* 79, no. 5 (2008): 1311–29.

44. Jerry Andriessen, "Arguing to Learn," in Sawyer, *Cambridge Handbook of the Learning Sciences*, 443–60; P. Bell, "Promoting Students' Argument Construction and Collaborative Debate in the Science Classroom," in *Internet Environments for Science Education*, ed. M. C. Linn, E. A. Davis, and P. Bell (Mahwah, NJ: Lawrence Erlbaum, 2004); Kuhn et al., "Arguing on the Computer"; T. Koschmann, "CSCL, Argumentation, and Deweyan Inquiry: Argumentation Is Learning," in *Arguing to Learn: Confronting Cognitions in Computer-Supported Collaborative Learning Environments*, ed. J. Andriessen, M. Baker, and D. Suthers (Dordrecht: Kluwer, 2003); Mark Klein, "The Deliberatorium: Crowd-Based Innovation," http://cci.mit.edu/klein/deliberatorium.html (accessed May 8, 2015).

45. Schrier, "Using Augmented Reality Games" and "Reliving History with *Reliving the Revolution*."

46. K. Schrier, J. Diamond, and D. Langendoen, "Using *Mission US: For Crown or Colony?* to Develop Historical Empathy and Nurture Ethical Thinking," in *Ethics and Game Design: Teaching Values through Play*, ed. K. Schrier and D. Gibson (Hershey, PA: IGI, 2010). For the game, which I helped design and write, see *Mission US* (Channel 13 / PBS, Electric Funstuff), www.mission-us.org (accessed May 6, 2015).

47. Nussbaum, "Collaborative Discourse, Argumentation, and Learning," 352; Matthew Keefer, Colleen M. Zeitz, and Lauren B. Resnick, "Judging the Quality of Peer-Led Student Dialogues," *Cognition and Instruction* 18, no. 1 (2000): 53–81.

48. T. Choontanom and B. Nardi, "Theorycrafting: The Art and Science of Using Numbers to Interpret the World," in Steinkuehler, Squire, and Barab, *Games, Learning, and Society*.

49. Michele M. Dornisch and Susan M. Land, "A Conceptual Framework for the Integration of Multiple Perspectives with Distributed Learning Environments," *Journal of Computing in Higher Education* 14, no. 1 (2002): 3–27.

50. Squire, *Video Games and Learning*.

51. "To achieve full citizenship, our children need to work with simulations that teach about the nature of simulation itself." Sherry Turkle, *The Second Self: Computers and the Human Spirit* (Cambridge, MA: MIT Press, 2005), 13.

52. Gee, *What Games Have to Teach Us*, 54, 66.

53. Ibid.

54. Ibid.

55. Barry Bozeman, Daniel Fay, and Catherine P. Slade, "Research Collaboration in Universities and Academic Entrepreneurship: The-State-of-the-Art," *Journal of Technology Transfer* 38 (2013): 1–67, quote at 37.

56. Klein, "Enabling Large-Scale Deliberation."

Chapter 6 · Amateurs

1. "20 Million Observations," www.aavso.org/20-million-observations; "20-Million Milestone for 100-Year Citizen Science Project," www.aavso.org/20-million-milestone-100-year -citizen-science-project (both accessed Oct. 30, 2013). A variable star is one that changes its brightness over time. According to the AAVSO director Arne Henden, "Because some variable stars are unpredictable and/or change their brightness over long time scales, it is not practical for professional astronomers to watch them every night. Thus, amateurs were recruited to keep tabs on these stars on behalf of professionals." GV Andromeda is "a member

of a class of older, pulsating stars smaller than our Sun." "20-Million Milestone." This donated work has amounted to a fully searchable and open database, which according to the AAVSO, is queried hundreds of times per day by researchers. A charge-coupled device (CCD) camera can interpret light waves and then convert it to data that can be recorded and interpreted by a computer. In this case, the camera is used to capture the light from a star and convert it to an electrical charge; the intensity of the charge relates back to the color it is reading from the star. For more about CCD cameras, see www.specinst.com/What_Is_A_CCD.html.

2. Michael A. Nielsen, *Reinventing Discovery: The New Era of Networked Science* (Princeton, NJ: Princeton University Press, 2012), 5; www.galaxyzoo.org/#/story (accessed Nov. 23, 2013). For Radio Galaxy Zoo, see http://radio.galaxyzoo.org (accessed Feb. 28, 2015).

3. Nielsen, *Reinventing Discovery*, 5. The number of papers listed at www.galaxyzoo.org /#/papers doubled between Nov. 23, 2013 (25 papers) and Feb. 28, 2015 (56 papers).

4. Janis L. Dickinson and Rick Bonney, *Citizen Science: Public Participation in Environmental Research* (Ithaca, NY: Comstock, 2012), 1. One issue to note is that while participation in amateur scientific research has persisted, scientists have not consistently encouraged contributions from amateurs in most areas of science, except in the areas of bird watching and astronomy. For more on citizen science projects, see chapter 1.

5. Dickinson and Bonney, *Citizen Science*, 19.

6. Ibid., 4. Dickinson and Bonney explain, "Volunteer bird surveys began in Europe in the eighteenth century, and North American lighthouse keepers began collecting data about bird strikes in 1880" (4). See also http://birds.audubon.org/how-christmas-bird-count-helps -birds and http://feederwatch.org/about/project-overview/#history-of-feederwatch (accessed Nov. 23, 2013). The 2012–13 season involved more than 20,000 FeederWatch participants.

7. Elizabeth Keeney, *The Botanizers* (Chapel Hill: University of North Carolina Press, 1992), 22–23.

8. Forrest M. Mims III, "Amateur Science—Strong Tradition, Bright Future" *Science* 284. no. 5411 (1999): 55–56. In regard to the amateur scientists, on a practical level, few people could make a living being a scientist, particularly before the twentieth century, so they had to do it on the side as amateurs. According to Steven Shapin, "In Britain alone, the list of amateur-scientists in the late eighteenth and nineteenth century includes some of the most influential figures in all the sciences." Shapin, *The Scientific Life: A Moral History of a Late Modern Vocation* (Chicago: University of Chicago Press, 1992), 42.

9. Dickinson and Bonney, *Citizen Science*; Nielsen, *Reinventing Discovery*.

10. Richard Louv, preface to Dickinson and Bonney, *Citizen Science*, ix; Keeney, *Botanizers*. Specifically, Keeney explains, "it did not purport of 'one who cultivates an activity as a pastime' until the early nineteenth century: the corollary imputations of inferior status to the pursuer of a pastime came still later, somewhere in mid-century" (4).

11. Shapin, *Scientific Life*, 46. See also a quote by Max Weber (1922): "Science today is a 'vocation' organized in special disciplines in the service of self-clarification and knowledge of interrelated facts. It is not the gift of grace of seers and prophets dispensing sacred values and revelations, nor does it partake of the contemplation of sages and philosophers about the meaning of the universe." Ibid., 152, citing Max Weber, "Science as a Vocation," from *Essays in Sociology* (New York: Routledge, 2009).

12. Shapin, *Scientific Life*, 57; Nielsen, *Reinventing Discovery*

13. Nielsen, *Reinventing Discovery*, 175.

14. Thorstein Veblen, *The Theory of the Leisure Class* (1899), www.gutenberg.org/files /833/833-h/833-h.htm (accessed Oct. 31, 2013).

15. See Lion Hirth, "State, Cartels and Growth: The German Chemical Industry," 2007, GRW Verlag Research Paper, www.grin.com/en/e-book/78269/state-cartels-and-growth-the -german-chemical-industry, for a much more nuanced look at this simplified example.

16. Elizabeth Eisenstein, *The Printing Press as an Agent of Change* (Cambridge: Cambridge University Press, 1979), 462, 487. "Learned journals," explains Eisenstein, "did contribute significantly to a sharper definition of the professional scientist, to new divisions of intellectual labor and to the creation of a 'referee system' . . . But it did not usher in serial publication, the preservation of data, the 'shift from motivated secrecy to motivated public disclosure,' or a feedback from informed readers to responsive authors and editors" (462). While Eisenstein posits that "print culture" actively shaped modern science with elements like fixity, Adrian Johns argues that these elements of so-called print culture need to be understood within culture—print culture was "because of culture just as much as causing culture." Johns, *The Nature of the Book* (Chicago: University of Chicago Press, 1998), 20.

17. Nielsen, *Reinventing Discovery*, 174–75.

18. Eisenstein, *Printing Press*, 487.

19. Ibid., 487–88.

20. Ibid.

21. Ibid., 477.

22. D. McLean and A. Hurd, *Kraus' Recreation and Leisure in Modern Society*, 9th ed. (Sudbury, MA: Jones and Bartlett Learning, 2011), 63. The Protestant work ethic came out of the Protestant Reformation of the 1500s. For more information about this development, see N. P. Miller and D. M. Robinson, *The Leisure Age: Its Challenge to Recreation* (Belmont, CA: Wadsworth Publishing Company, 1963), and their description of "purity of conduct." Cited in McLean and Hurd, *Kraus' Recreation and Leisure*, 66. Puritans who arrived in New England adopted this work ethic, denigrating idleness as the "devil's workshop." Policies at that time even forbade "merrymaking," such as gambling and dancing. See a more nuanced account of Puritan leisure in B. C. Daniels, "Did the Puritans Have Fun? Leisure, Recreation, and the Concept of Pleasure in Early New England," *Journal of American Studies* 25, no. 1 (1991): 7–22.

23. McLean and Hurd, *Kraus' Recreation and Leisure*.

24. Ibid., 83; Verblen, *Theory of the Leisure Class*. On hierarchies of taste, see, e.g., Pierre Bourdieu, *Distinction: A Social Critique of the Judgement of Taste*, trans. Richard Nice (Cambridge, MA: Harvard University Press, 1984).

25. Keeney, *Botanizers*, 95.

26. Max Weber, "On Science as a Vocation," 1917, www.wisdom.weizmann.ac.il/~oded/X /WeberScienceVocation.pdf (accessed Nov. 27, 2013).

27. See, e.g., how Erving Goffman describes framing. "Goffman describes social worlds as constituting frames of experience. He defines a frame as a situational definition constructed in accord with organizing principles that govern both the events themselves and participants' experiences of these events." G. A. Fine, "Frames and Games," in *The Game Design Reader*, ed. Katie Salen and Eric Zimmerman (Cambridge, MA: MIT Press, 2006), 579.

28. Ibid., 581. Bernard Suits explains that rules are obeyed because "such obedience is a necessary condition for my engaging in the activity such obedience makes possible." Suits, "Construction of a Definition," in Salen and Zimmerman, *Game Design Reader*, 181. For more on frame analysis, see Erving Goffman, *Frame Analysis: An Essay on the Organization of Experience* (Boston, MA: Northeastern University Press, 1974).

29. "Rhetoric of play as frivolous is usually applied to the activities of the idle or the foolish. But in modern times, it inverts the classic 'work ethic' view of play, against which all the other rhetorics exist as rhetorics of rebuttal. But frivolity, as used here, is not just the puritanic negative, it is also a term to be applied more to historical trickster figures and fools who were once the central and carnivalesque persons who enacted playful protest against the orders of the ordained world." Brian Sutton-Smith, "Play and Ambiguity," in Salen and Zimmerman, *Game Design Reader*, 305–6.

30. Katie Salen and Eric Zimmerman, *Rules of Play* (Cambridge, MA: MIT Press, 2003); Amy Bruckman, "Can Educational Be Fun?" www.cc.gatech.edu/~asb/papers/bruckman -gdc99.pdf; Ian Bogost, "Fun," UX Week 2013 talk, http://vimeo.com/74943170 (both accessed July 2, 2015); Eric Zimmerman, "Play as Research: The Iterative Design Process," in *Design Research: Methods and Perspectives*, ed. Brenda Laurel (Cambridge, MA: MIT Press, 2003). On making reality more fun, see, e.g., Jane McGonigal, *Reality Is Broken* (New York: Penguin, 2011).

31. See Jesper Juul, *The Art of Failure: An Essay on the Pain of Playing Video Games* (Cambridge, MA: MIT Press, 2013).

32. Bogost, "Fun."

33. Fine, "Frames and Games," 593.

34. S. Sniderman, "Unwritten Rules," in Salen and Zimmerman, *Game Design Reader*, 502.

35. Bogost, "Fun"; Suits, "Construction of a Definition," in Salen and Zimmerman, *Game Design Reader*, 190.

36. This is a play on the phrase "half-real" from Juul, "Half-Real."

37. "Lulz" and "Did it for the 'lulz'" references a meme on the online bulletin board 4Chan, which was discussed vividly in Julian Dibbell's "Radical Opacity," published in MIT's *Technology Review*, Aug. 23, 2010, www.technology.review.com/featuredstory/420323 /radical-opacity/ (accessed Aug. 1, 2015). He explains that the lulz are "laughs, jest, cheap amusement, but in a broader sense encompasses both the furious creativity that generates /b/'s vast repertoire of memes and the rollicking subcultural intensity they inspire." /b/ is a random topics section in an online forum or subtopic (subreddit) on Reddit.

38. "Knowledge seems to be inescapably social: it depends on the face-to-face interactions that help constitute such a culture." Johns, *Nature of the Book*, 45.

39. On the shift from oral to print culture, see Walter J. Ong, *Orality and Literacy: The Technologizing of the Word* (New York: Routledge, 1982). Prior to printed books, people would look at the structure of one's script in a handwritten text to evaluate whether it was a trustworthy source.

40. Johns, *Nature of the Book*, 31. Steven Shapin explains that the "identification of trust as a key element in the making of knowledge." Shapin, *A Social History of Truth: Civility and Science in Seventeenth-Century England* (Chicago: University of Chicago Press, 1995), 129.

41. Shapin, *Social History of Truth*; Johns, *Nature of the Book*.

42. Higgins, *Beyond Pleasure and Pain: How Motivation Works* (New York: Oxford University Press, 2012).

43. See, e.g., recent articles on fraudulent research or issues replicating results, such as Alok Jha, "False Positives: Fraud and Misconduct Are Threatening Scientific Research," *Guardian*, Sept. 13, 2012, www.theguardian.com/science/2012/sep/13/scientific-research-fraud -bad-practice (accessed July 8, 2015). See also issues of gender bias and racial discrimination in peer review, such as #AddMaleAuthorGate and Silvia Knobloch-Westerwick, Carroll J. Glynn, and Michael Huge, "The Matilda Effect in Science Communication: An Experiment on Gender Bias in Publication Quality Perceptions and Collaboration Interest," *Science Communication* 35 (2013): 603–25.

44. Mark B. Brown, *Science in Democracy* (Cambridge, MA: MIT Press, 2009), 13, 171, citing Bruno Latour, *Pandora's Hope* (Cambridge, MA: Harvard University Press, 1999). On scientific research, see Shapin, *Scientific Life*.

45. On the mixing of science, technology, and politics see, e.g., Bruno Latour, *Science in Action* (Cambridge, MA: Harvard University Press, 1987).

46. Brown, *Science in Democracy*, 171.

47. James Vincent, "MIT's 'Immersion' Project Reveals the Power of Metadata," *Independent*, July 8, 2013, www.independent.co.uk/life-style/gadgets-and-tech/mits-immersion-project-reveals-the-power-of-metadata-8695195.html (accessed July 8, 2015).

Chapter 7 · Participation

1. *Happy Moths* (Kevin Crowston, Syracuse University iSchool), http://socs.ischool.syr.edu/happymatch/index.php/GameInitialization/startGame/59 (accessed Feb. 26, 2015).

2. Jane McGonigal, *Reality Is Broken* (New York: Penguin, 2011), 333–34. An ARG combines different platforms (real world locations or events, social media, video, online games, telephone) to tell a story and provide missions and puzzles to participants over time. Typically, participants can play together in person or virtually, and share secrets, solutions, and hints online via forums and other communication platforms. Bogost explains that Jane McGonigal defines them as *"any* game that integrates itself with the real world, not just one that involves the usual trappings of that genre, like distributed narrative and puzzle-solving." Bogost, "Reality Is Alright: A Review of Jane McGonigal's Book *Reality Is Broken*," Jan. 14, 2011, http://bogost.com/blog/reality_is_broken (accessed Mar. 1, 2015). For more on *Urgent Evoke* (World Bank Institute, Jane McGonigal), see www.urgentevoke.com/page/mission-list and Lesley Bauer, "A Look at World Bank's *Urgent Evoke*," http://techchange.org/2011/03/18/a-look-at-urgent-evoke-reflections-for-season-2 (accessed Mar. 1, 2015).

3. Bauer, "A Look at World Bank's *Urgent Evoke*."

4. Yuxiang Zhao and Qinghua Zhu, "Evaluation on Crowdsourcing Research: Current Status and Future Direction," *Information Systems Frontiers* 16 (2014): 417–34, 427; M. Jordan Raddick et al., "Galaxy Zoo: Motivations of Citizen Scientists," *Astronomy Education Review* 9 (2010): 31.

5. Osamuyimen Stewart, David Lubensky, Juan M. Huera, "Crowdsourcing Participation Inequality: A SCOUT Model for the Enterprise Domain," in *Proceedings of the ACM SIGKDD Workshop on Human Computation* (New York: ACM, 2010), 30. The authors identified three different types of participation, *collectivistic* (with people each working together to solve small parts of a larger problem), *individualistic* (many people contributing individually), and *collectivistic-individualistic* (which mixes the two, or groups who can collaborate with each other but compete against other groups). Their study focused on only the individualistic type of participation, and the results suggest that supercontributors are typically driven by an intrinsic reward, such as wanting to contribute, and contributors are motivated by extrinsic rewards, such as badges or money. Outliers may have an interest in the project, but it is not sufficient to encourage sufficient effort in it. The researchers were able to further participation by providing differential incentives (multitiered incentives).

6. Oded Nov, Ofer Arazy, and David Anderson, "Scientists@Home: What Drives the Quantity and Quality of Online Citizen Science Participation," *PLoS ONE* 9, no. 4 (2014): e90375. *Stardust@home* (UC Berkeley Space Sciences Lab), http://stardustathome.ssl.berkeley.edu, is a project where participants are invited to classify and categorize images from the NASA Stardust spacecraft in an effort to find tracks left by space particles. Nov, Arazy, and Anderson also looked at the Citizen Weather Observer Program (CWOP) and Berkeley Open Infrastructure for Network Computing (BOINC).

7. Katherine Xue, "Popular Science: In the Internet Era, Research Moves from Professionals' Labs to Amateurs' Homes," *Harvard Magazine*, Jan.–Mar. 2014, http://harvardmagazine.com/2014/01/popular-science; ESA, "Essential Facts about the Computer and Video Game Industry," 2014, www.theesa.com/wp-content/uploads/2014/10/ESA_EF_2014.pdf (both accessed May 3, 2015).

8. Neil Seeman, "Don't Mistake Likes on Facebook for Real Social Change," *Huffington Post*, Nov. 19, 2013 www.huffingtonpost.ca/neil-seeman/wisdom-of-crowds_b_4297913.html. Thunderclap aims to help people amplify their messages on social media; see www.thunderclap .it/about (both accessed May 3, 2015).

9. John Wilson, "Volunteerism Research: A Review Essay," *Nonprofit and Voluntary Sector Quarterly* 41, no. 2 (2012): 186, 187; Samuel Shye, "The Motivation to Volunteer: A Systemic Quality of Life Theory," *Social Indicators Research* 98 (2010): 183–200.

10. Wilson, "Volunteerism Research," 186–87.

11. Daren Brabham, "Moving the Crowd at iStockphoto: The Composition of the Crowd and Motivations for Participation in a Crowdsourcing Application," *First Monday* 13, no. 6 (2008), http://firstmonday.org/article/view/2159/1969 (accessed Feb. 25, 2012); Raddick et al., "Galaxy Zoo: Motivations of Citizen Scientists"; Karim R. Lakhani, Lars Bo Jeppesen, Peter A. Lohse, and Jill A. Panetta, "The Value of Openness in Scientific Problem Solving," working paper, 2007, www.hbs.edu/faculty/publication%20files/07-050.pdf (accessed Aug. 1, 2015).

12. ESA, "2015 Essential Facts about the Computer and Video Game Industry," www .theesa.com/wp-content/uploads/2015/04/ESA-Essential-Facts-2015.pdf (accessed Sept. 7, 2015).

13. *Astro Drone Crowdsourcing Game* (Advanced Concepts Team, European Space Agency; Micro Air Vehicle Laboratory, TU Delft; Artificial Intelligence Group, Radboud University Nijmegen), www.esa.int/gsp/ACT/ai/projects/astrodrone.html (accessed Mar. 1, 2015).

14. Henry Jenkins, Ravi Purushotma, Katie Clinton, and Alice J. Robison, "Confronting the Challenges of Participatory Culture: Media Education for the 21st Century," The John D. and Catherine T. MacArthur Foundation Reports on Digital Media and Learning (Cambridge, MA: MIT Press, 2009); Eric Zimmerman, "Gaming Literacy: Game Design as a Model for Literacy in the Twenty-First Century," in *The Video Game Theory Reader 2*, ed. Bernard Perron and Mark J. P. Wolf (New York: Routledge, 2009); Matthew J. Gaydos and Kurt D. Squire, "Role Playing Games for Scientific Citizenship," *Cultural Studies of Science Education* 7 (2012): 821–44.

15. "Popular magazines, too, have long understood the value of reader contributions: one of Australia's culturally and politically formative 19th-century magazines, *The Bulletin*, cherished for many years the tagline, 'half Australia writes it, all Australia reads it' . . . [User-generated content] also has an important role in highly regulated 20th-century electronic media, notably in programming formats such as talk-back radio . . . as well as in open-access and community radio and television channels . . . More recently, websites seeking user content for the purposes of a commercial promotion—'Invent our new flavor!,' 'Caption this photo/cartoon,' and so on—generate carefully managed, legally controlled transactions soliciting user involvement in highly formalized environments." Ramon Lobato, Julian Thomas, and Dan Hunter, "Histories of User-Generated Content: Between Formal and Informal Media Economies," *International Journal of Communication* 5 (2011): 903.

16. Bruns, "Produsage: Towards a Broader Framework for User-Led Content Creation," in *Proceedings of the 6th ACM SIGCHI Conference on Creativity & Cognition: Washington, D.C., June 13–15, 2007* (New York: ACM, 2007), 99.

17. Christian Fuchs, "Digital Prosumption Labour on Social Media in the Context of the Capitalist Regime of Time," *Time & Society* 23, no. 1 (2014): 97–123; John Banks and Jason Potts, "Co-creating games: a Co-Evolutionary Analysis," *New Media & Society* 12, no. 2 (2010): 253–70. On the prosumer, see also Susan Gunelius, "The Shift from CONsumer to PROsumer," *Forbes*, July 30, 2010, www.forbes.com/sites/work-in-progress/2010/07/03/the -shift-from-consumers-to-prosumers (accessed May 3, 2015).

18. See, e.g. a forum on modifications for the video game *Minecraft* (Mojang), www .curse.com/mc-mods/minecraft/railcraft (accessed Feb. 28, 2015).

19. Justin McElroy, "Wright: 'Spore Fans 38% God,'" *Engadget* www.joystiq.com/2008/07 /14/wright-spore-fans-38-god (accessed May 4, 2015). See the species users created for *Spore* (Electronic Arts), at *Sporepedia*, www.spore.com/sporepedia (accessed Mar. 1, 2015).

20. Jennifer L. Shirk et al., "Public Participation in Scientific Research: A Framework for Deliberate Design," *Ecology and Society* 17, no. 2 (2012): 29.

21. Yochai Benkler, *The Wealth of Networks: How Social Production Transforms Markets and Freedom* (New Haven, CT: Yale University Press, 2006); John Banks and Sal Humphreys, "The Labour of User Co-Creators," *Convergence* 14, no. 4 (2008): 405–6; Banks and Potts, "Co-creating Games." Banks and Humphreys explain, "Benkler develops an argument that distributed peer production networks, which create *Wikipedia* and open source projects such as Linux, may well be more effective, efficient and sustainable than firmbased market activity. He works from a transaction costs analysis to propose that distributed peer networks are particularly effective and efficient at allocating scarce resources such as 'human creativity, time, and attention' (2006: 107). Human creative labour is notoriously difficult to standardize, specify, price and then organize as an input cost. Firms and monetized markets, however, rely on such contractual specification. The social sharing of non-market peer-production on the other hand does not require or rely on such precise specification (Benkler, 2006: 110–16)." Banks and Humphreys, "Labour of User Co-Creators," 405–6, citing Benkler, *Wealth of Networks*.

22. Lawrence Lessig, *Code 2.0* (New York: Basic Books, 2006), 276, cited in Christian Fuchs, "Book Review: Don Tapscott and Anthony D. Williams, *Wikinomics*," *International Journal of Communication* 2 (2008): 1–11, 9. For an example of a game that explores the idea of property, see *The Free Culture Game* (Mollendustria), www.molleindustria.org/en /freeculturegame (accessed May 3, 2015).

23. Jenkins et al., "Confronting the Challenges of Participatory Culture."

24. Jean Lave and Etienne Wenger, *Situated Learning: Legitimate Peripheral Participation* (New York: Cambridge University Press, 1991). According to Jose P. Zagal and Amy Bruckman, "A community of practice involves a collection of individuals sharing mutually defined practices, beliefs, and understandings over an extended time frame in the pursuit of a shared enterprise." Zagal and Bruckman, "Designing Online Environments for Expert/ Novice Collaboration: Wikis to Support Legitimate Peripheral Participation," *Convergence* 16, no. 4 (2010): 451–70, 452.

25. Zagal and Bruckman, "Designing Online Environments," 452, citing Lave and Wenger, *Situated Learning.*

26. James P. Gee and Elisabeth Hayes, "Nurturing Affinity Spaces and Game-Based Learning," in *Games, Learning, and Society: Learning and Meaning in the Digital Age*, ed. Constance Steinkuehler, Kurt Squire, and Sasha Barab, 129–53 (New York: Cambridge University Press, 2012); Kurt Squire, *Video Games and Learning: Teaching and Participatory Culture in the Digital Age* (New York: Teachers College Press, 2011); Constance Steinkuehler and Sean Duncan, "Scientific Habits of Mind in Virtual Worlds," *Journal of Science Education and Technology* 17 (2008): 530–43; Constance Steinkuehler and Yoonsin Oh, "Apprenticeship in Massively Multiplayer Online Games," in Steinkuehler, Squire, and Barab, *Games, Learning, and Society.*

27. Gee and Hayes, "Nurturing Affinity Spaces."

28. See ibid. In Gee's affinity group definition, people are bonded through common interests with extensive and intensive knowledge; flexible roles and functions; tacit, dispersed, and distributed knowledge; and leaders who help design and learn from the group. James Gee, *What Games Have to Teach Us about Learning and Literacy* (New York: Palgrave Macmillan, 2003), 193.

29. Lave and Wenger, *Situated Learning*; Zagal and Bruckman, "Designing Online Environments," 452. You could replace "technology or platform" with "game" and it still applies. See also Gee, *What Games Have to Teach Us*, and Salen and Zimmerman, *Game Design Reader*.

30. Jenkins et al., "Confronting the Challenges of Participatory Culture." Jenkins explains that transmedia worlds are ones that exist across multiple platforms, such as a website, book, and video game, where aspects of the world are explored through each platform. The world itself is cohesive, but each individual story or platform is also able to provide a satisfying experience. "Transmedia storytelling represents a process where integral elements of a fiction get dispersed systematically across multiple delivery channels for the purpose of creating a unified and coordinated entertainment experience. Ideally, each medium makes its own unique contribution to the unfolding of the story." Henry Jenkins, "Transmedia 202: Further Reflections," *Confessions of an Aca-Fan*, Aug. 1, 2011, http://henryjenkins.org/2011 /08/defining_transmedia_further_re.html (accessed May 3, 2015). The 39 Clues (Scholastic) is a series of books, games, and websites for kids that involves a mystery and a series of clues that lead to real-world prizes. See http://the39clues.scholastic.com (accessed May 3, 2015).

31. "As Marx articulates in *Capital* (1976), it is through the performance of labor that we come to realize our own potential." J. E. Campbell, in Mark Andrejevic, John Banks, John Edward Campbell, Nick Couldry, Adam Fish, Alison Hearn, and Laurie Ouellette, "Participations: Dialogues on the Participatory Promise of Contemporary Culture and Politics, Part 2: Labor," *International Journal of Communication* 8 (2014): 1089–106, 1096, citing Karl Marx, *Capital: A Critique of Political Economy*, trans. Ben Fowkes (Harmondsworth, UK: Penguin, 1976). Campbell also recalls that "in *Grundrisse* (1973), Marx sees labor as a process of discovery; through labor individuals discover aspects of themselves (talents, ambitions, needs, desires) of which they were previously unaware. They grow not only in regard to their skills and abilities, but also in terms of their self-conceptions" (1097, citing Marx, *Grundrisse*, trans. David McLellan [St. Albans, UK: Paladin, 1973]).

32. Tiziana Terranova, "Free Labor: Producing Culture for the Digital Economy," *Social Text* 18, no. 2 (2000), quotes at 37 and 38. In regard to knowledge labor, Christian Fuchs argues that "knowledge labour is labour that produces and distributes information, communication, social relationships, affects, and information and communication technologies." Fuchs, "Class, Knowledge, and New Media," *Media, Culture, & Society* 32, no. 1 (2010): 141.

33. Terranova bases this assumption on Lazzarato's definition that immaterial labor, which "involves a series of activities that are not normally recognized as 'work' . . . such as the kinds of activities involved in defining and fixing cultural and artistic standards, fashions, tastes, consumer norms, and, more strategically, public opinion." Terranova, "Free Labor," 41, quoting Maurizio Lazzarato, "Immaterial Labor," in *Marxism beyond Marxism*, ed. Saree Makdisi, Cesare Casarino, and Rebecca E. Karl for the Polygraph Collective (London: Routledge, 1996), 133. See also Fuchs, "Class, Knowledge, and New Media."

34. "This affective, immaterial labour is neither directly produced by capital, nor developed as a direct response to the needs of capital. This process should not be understood as a straightforward incorporation or appropriation of the free labour of an otherwise authentic fan culture. This free labour has not been appropriated but voluntarily given. The relationships are much more nuanced and complex than the language of incorporation, appropriation or exploitation suggests. Rather, Terranova proposes that the dynamics reconfiguring relations between production and consumption are played out within a field that 'is always and already capitalism'; they are immanent to the networks of informational capitalism (Terranova, 2004: 79)." Banks and Humphreys, "Labour of User Co-Creators," 407, citing Terranova, "Free Labor," and Terranova, *Network Culture: Politics for the Information Age* (London: Pluto Press, 2004), 79.

35. Terranova, "Free Labor," 48.

36. Fuchs, "Class, Knowledge, and New Media," 141.

37. On defining passionate labor, see Hector Postigo, "America Online Volunteers: Lessons from an Early Co-Production Community," *International Journal of Cultural Studies* 12, no. 5 (2009): 451–69. Campbell explains that passionate labor "provide[s] sufficient emotional returns that they aren't even recognized as work by those performing them." Campbell, in Andrejevic et al., "Participations," 1097.

38. Machinima uses real-time video game engines to create mini-movies.

39. "The informal media economy encompasses an extremely diverse range of production activities—including DIY publishing, slash video, and many other forms of amateur production—along with an equally large range of distribution activities, from disc piracy and peer-to-peer file-sharing through to second-hand markets and the parallel-importation of CDs, DVDs, and games." Lobato, Thomas, and Hunter, "Histories of User-Generated Content," 5.

40. Terranova, "Free Labor"; Lev Grossman, "You—Yes, You—Are TIME's Person of the Year," *Time Magazine*, Dec. 25, 2006, http://content.time.com/magazine/article/0,9171,1570810,00 .html (accessed May 4, 2015), cited in Banks and Humphreys, "Labour of User Co-Creators," 401.

41. "Much of the industry interest in the proposed Spore approach to content creation concerns the potential to address rapidly escalating development costs in the videogames industry, particularly associated with the creation of art content. Wright is upfront on this as an exercise in cost reduction, in essentially outsourcing labour to this collective intelligence, peer production network." Banks and Humphreys, "Labour of User Co-Creators," 404.

42. Hector Postigo, "The Socio-Technical Architecture of Digital Labor: Converting Play into YouTube Money," *New Media Society* (2014): 3; Sut Jhally, *The Codes of Advertising* (New York: Routledge, 1987); Dallas Smythe, "On the Audience Commodity and Its Work" (1981), reprinted in *Media and Cultural Studies KeyWorks*, ed. Meenakshi Gigi Durham and Douglas M. Kellner (Malden, MA: Blackwell, 2001), 256, 261; Fuchs, "Class, Knowledge, and New Media," 148.

43. "The difference between the audience commodity on traditional mass media and on the Internet is that in the latter the users are also content producers; there is user-generated content, the users engage in permanent creative activity, communication, community building and content-production." Fuchs, "Class, Knowledge, and New Media," 148.

44. "Kücklich, for example, suggests that industries tout a rhetoric of collaboration which masks their profit-seeking motives—assuming perhaps that players do not understand the value of their contributions—(Kücklich, 2005). In a recent article, Herman, Coombe and Kaye conclude that playing in Second Life may well be a 'half-life' of 'corporate servitude' in which participants misrecognize their social relations as part of an intellectual property exchange (Herman et al., 2006: 204). The underlying assumption of these discourses is that players are in some sense unaware that their participation is a productive practice from which economic value is extracted. Here labour is deployed as a category not so much to describe or deploy emerging modes of agency and associations between actors. Instead it becomes an explanation in the form of critique, which seeks to unveil or disclose social forces that may be at work behind the actors' backs (Latour, 2005: 136)." Banks and Humphreys, "Labour of User Co-Creators," 404–5, citing Julian Kücklich, "Precarious Playbour: Modders and the Digital Games Industry," *Fibreculture Journal* 5 (2005), http://five .fibreculturejournal.org/fcj-025-precarious-playbour-modders-and-the-digital-games -industry/; A. Herman, R. J. Coombe, and L. Kaye, "Your Second Life? Goodwill and the Performativity of Intellectual Property in Online Digital Gaming," *Cultural Studies* 20, no. 2–3 (2006): 184–210; and Bruno Latour, *Reassembling the Social: An Introduction to Actor-Network Theory* (Oxford: Oxford University Press, 2005).

45. Postigo, "Socio-Technical Architecture of Digital Labor," 10.

46. Fuchs, "Digital Prosumption Labour," 120. "Turning leisure time into labour time is one attempt at prolonging capitalism . . . More disposable time means more time for consumption, creativity and leisure, to which capital is connected" (120–21). As a result, the boundaries between "play and labour, work time and leisure time, production and consumption, the factory and the household, public and private life tend to blur" (110). See also Fuchs, "Class, Knowledge, and New Media."

47. "When we convert leisure time into labour time, . . . labour goes beyond the factory and seeps into the home and leisure time." Fuchs, "Digital Prosumption Labour," 112. See also Fuchs, "Class, Knowledge, and New Media."

48. Terranova, "Free Labor," 48.

49. See, e.g., ea_spouse at Erin Hoffman, "EA: The Human Story," http://ea-spouse .livejournal.com/274.html (accessed May 4, 2015).

50. Fuchs, "Class, Knowledge, and New Media," referencing Marx, *Capital*. In Marx's distinction, the exploited class is separated from the means of production, so that there is a separation between workers and the ownership of that work, explains Fuchs.

51. Andrejevic, in Andrejevic et al., "Participations," 1090–91, citing Nancy Holmstrom "Exploitation," in *Exploitation: Key Concepts in Critical Theory*, ed. K. Nielsen and R. Ware (New York: Humanities Press International, 1997), 81–102. Andrejevic characterizes exploit-ative activities as transforming "our own activity (or at least an important part of it) back upon ourselves in unrecognizable form, serving interests and imperatives that are not our own." Andrejevic, in Andrejevic et al., "Participations," 1091. He also recalls Erik Olin Wright, as paraphrased by David Hesmondhalgh, "User-Generated Content, Free Labour, and the Cultural Industries," *Ephemera: Theory and Politics in Organization* 10, no. 3–4 (2010): 267–84, 274, explaining that "exploitation occurs when the material welfare of one class is causally dependent upon the material deprivation of another . . . Second, that causal dependence depends in turn on the exclusion of workers from key productive resources, especially property. Third, the mechanism through which both these features (causal dependence and exclusion) operate is appropriation of the labour of the exploited" (1090).

52. Fuchs, "Book Review," 3, 6, for direct quotations in this paragraph. On precarious labor, see also Fuchs, "Digital Prosumption Labour," 99, where his argument draws from Marxist conceptions of labor and critiques of capitalism. Produser or prosumer does not signal democratization of the media, but it is "total commodification of human creativity." Fuchs, "Class, Knowledge, and New Media," 149. Fuchs also explains that not all people will benefit from a "Wikinomic" type model, and some people will benefit by taking advantage of free laborers. "Book Review," 6.

53. Andrew Ross, "The Mental Labor Problem," *Social Text* 18, no. 2 (2000): 1–31, cited in Andrejevic et al., "Participations," 1092; Banks and Humphreys, "Labour of User Co-Creators." Additionally, Banks explains that "sometimes apparently the same practice, under certain conditions and at something of a threshold, can start to be viewed as unfair, unreasonable, and even exploitative and coercive." In Andrejevic et al., "Participations," 1092.

54. "We have sought to move away from the discursive construction of user-creators as unknowing and exploited people who do not recognize the conditions under which they produce value. We want to take seriously their own, often sophisticated understanding of negotiations with enterprise, and their decision-making in the directions of both commercial and non-commercial production." Banks and Humphreys, "Labour of User Co-Creators," 415.

55. Andrejevic et al, "Participations," 1091.

56. Couldry, in ibid., 1091.

57. Campbell argues that we need to distinguish between what he calls oppressive and abstract exploitation. "Oppressive forms of exploitation are readily apparent to the worker and have a direct and obvious impact on the material existence of the laborer, whether in the sweatshops in Bangladesh, the Foxconn plants in China, or among workers at Walmart in the United States who are not paid a livable wage. However, in abstract modes of exploitation, the worker may remain unaware of the wealth his or her activities generate for a small class of people. This would be the case with what I identify as the 'labor of devotion'—a form of work found in brand communities." In Andrejevic et al., "Participations," 1097, citing J. E. Campbell, "It Takes an iVillage: Gender, Labor, and Community in the Age of Television-Internet Convergence," *International Journal of Communication* 5 (2011): 492–510.

58. Campbell, in Andrejevic et al., "Participations," 1097, citing K. Kuehn and T. F. Corrigan, "Hope Labor: The Role of Employment Prospects in Online Social Production," *Political Economy of Communication* 1, no. 1 (2013): 9–25, and "Podcast and Dialogue with Alice Marwick and Brooke Duffy," *Culture Digitally*, Mar. 26, 2013, http://culturedigitally.org/2013/03/podcast-and-dialogue-with-alice-marwick-and-brooke-duffy (accessed July 9, 2015).

59. Campbell, discussing Terranova's research, in Andrejevic et al., "Participations," 1097.

60. Hearn, Fish, Banks, and Andrejevic in ibid., 1090–92, 1099–1100.

61. Walter Benjamin, "Der Autor als Produzent," in *Medienästhetische Schriften* (Frankfurt: Suhrkamp, 1934), 243, cited in Fuchs, "Book Review," 2; Henry Jenkins, *Convergence Culture: Where Old and New Media Collide* (New York: NYU Press, 2006).

62. Andrejevic et al., "Participations," 1098, citing M. Hindman, *The Myth of Digital Democracy* (Princeton, NJ: Princeton University Press, 2008).

63. Banks and Potts, "Co-creating Games," 259.

64. Postigo, "Socio-Technical Architecture of Digital Labor," 4. He continues: "Affordances come in two flavors: technological and social. Technological affordances have been described as the set of functions that a technology makes possible: telephones make long distance conversation possible, antibiotics make defeating infections possible, and YouTube allows people to 'broadcast' homemade amateur videos. Social affordances are the social structures that take shape in association with a given technical structure. Telephones allowed for increased socialization changed our perception of distance (the telegraph did that), reshaped market practices, and so on" (5).

65. Banks and Potts, "Co-creating Games," 259.

66. Ibid., 261; Banks and Humphreys, "Labour of User Co-Creators," 412.

Chapter 8 · Data

1. *Play to Cure: Genes in Space* (Cancer Research UK and Guerilla Tea), iOS, www.cancerresearchuk.org/support-us/play-to-cure-genes-in-space (accessed July 9, 2015); Charles Lees-Czerkawski, "Postmortem: Guerilla Tea's Play to Cure: Genes in Space," *Gamasutra*, Mar. 13, 2014, http://gamasutra.com/blogs/CharlesLeesCzerkawski/20140313/213086/Postmortem_Guerilla_Teas_Play_to_Cure_Genes_in_Space.php (accessed Feb. 26, 2015).

2. Lees-Czerkawski, "Postmortem."

3. Lisa Gitelman and Virginia Jackson, introduction to *"Raw Data" Is an Oxymoron*, ed. Lisa Gitelman (Cambridge, MA: MIT Press, 2013), 1; Chris Anderson, "The End of Theory: The Data Deluge Makes the Scientific Method Obsolete," *Wired,* June 2008, http://archive.wired.com/science/discoveries/magazine/16-07/pb_theory (accessed Feb. 26, 2015).

4. Tom Kalil, "Big Data Is a Big Deal," Mar. 29, 2012, www.whitehouse.gov/blog/2012/03/29/big-data-big-deal (accessed Feb. 25, 2015); Joanna Pearlstein, "Information Revolution: Big Data Has Arrived at an Almost Unimaginable Scale," *Wired,* Apr. 16, 2013, www.wired.com/2013/04/bigdata.

5. Kenneth Neil Cukier and Viktor Mayer-Schönberger, "The Rise of Big Data: How It's Changing the Way We Think About the World," *Foreign Affairs*, May–June 2013, www .foreignaffairs.com/articles/2013-04-03/rise-big-data; Edd Dumbill, "What Is Big Data?" *O'Reilly*, Jan. 11, 2012, http://strata.oreilly.com/2012/01/what-is-big-data.html (both accessed May 4, 2015). In a survey of Big Data research, Chen, Mao, and Liu explain that although the definition of Big Data adapts to different contexts, they define it as "datasets that could not be perceived, acquired, managed, and processed by traditional IT and software/hardware tools within a tolerable time." Min Chen, Shiwen Mao, and Yunhao Liu, "Big Data: A Survey," *Mobile Networking Applications* 19 (2014): 171–209, 173.

6. Andrej J. Zwitter and Amelia Hadfield, "Governing Big Data," *Politics and Governance* 2, no. 1 (2014): 1. "It is estimated that humanity accumulated 180 EB of data between the invention of writing and 2006. Between 2006 and 2011, the total grew ten times and reached 1,600 EB. This figure is now expected to grow fourfold approximately every 3 years." L. Floridi, "Big Data and their Epistemological Challenge," *Philosophy and Technology* 25, no. 4 (2012): 435–37, 435. Chen, Mao, and Liu, "Big Data: A Survey," relate that 1.8 ZB of data have been generated thus far in the world, which should double every four years. An exabit (EB) is 10^{18} bits and a zettabit (ZB) is 10^{21} bits.

7. R. Raley, "Dataveillance and Countervailance," in Gitelman, *"Raw Data" Is an Oxymoron*; David Lyon, *Electronic Eye: The Rise of Surveillance Society* (Minneapolis: University of Minnesota Press, 1994); Mark Andrejevic, "The Big Data Divide," *International Journal of Communication* 8 (2014): 1677; Cornelius Puschmann and Jean Burgess, "Metaphors of Big Data," *International Journal of Communication* 8 (2014): 1690–709, 1692, citing D. Rosenberg, "Data before the Fact," in Gitelman, *"Raw Data" Is an Oxymoron*, 15–40.

8. Andrejevic, "Big Data Divide," 1676.

9. Chen, Mao, and Liu, "Big Data: A Survey."

10. See more in ibid.

11. Puschmann and Burgess, "Metaphors of Big Data," 1702; danah boyd and Kate Crawford, "Critical Questions for Big Data: Provocations for a Cultural, Technological, and Scholarly Phenomenon," *Information, Communication, & Society* 15, no. 5 (2012): 662–79; Evelyn Ruppert, "Rethinking Empirical Social Sciences," *Dialogues in Human Geography* 3, no. 3 (2013): 268–73; Evelyn Ruppert, John Law, and Mike Savage, "Reassembling Social Science Methods: The Challenge of Digital Devices," *Theory, Culture & Society* 30, no. 4 (2013): 22–46.

12. Cukier and Mayer-Schönberger, "Rise of Big Data"; Andrejevic, "Big Data Divide," 1676. See also "Australians Concerned for Online Privacy," www.uq.edu.au/news/article /2012/03/australians-concerned-online-privacy (accessed July 9, 2015).

13. Cukier and Mayer-Schoenberger, "Rise of Big Data"; Anderson, "End of Theory," para. 13.

14. James Surowiecki, *The Wisdom of Crowds* (New York: Random House, 2005).

15. Andrejevic, "Big Data Divide," 1676, quoting D. Weinberger, *Too Big to Know: Rethinking Knowledge Now That the Facts Aren't the Facts, Experts Are Everywhere, and the Smartest Person in the Room Is the Room* (New York: Basic Books, 2011), 130. Andrejevic continues, "Data mining promises to generate patterns of actionable information that outstrip the reach of the unaided human brain."

16. Cukier and Mayer-Schoenberger, "Rise of Big Data," 6.

17. Cukier and Mayer-Schoenberger, "Rise of Big Data"; Puschmann and Burgess, "Metaphors of Big Data," 1691. On computer programming for historians, see Ching-man Au Yeung and Adam Jawowt, "Studying How the Past Is Remembered: Towards Computational

History through Large Scale Text Mining," presented at the International Conference on Conference on Information and Knowledge Management, Glasgow, Scotland, October 24–28, 2011, www.dl.kuis.kyoto-u.ac.jp/~adam/cikm11a.pdf, and the Programming Historian tutorials, http://programminghistorian.org (both accessed May 8, 2015).

18. Alex Pentland, *Social Physics: How Good Ideas Spread—The Lessons from a New Science* (New York: Penguin, 2014); Ben Waber, *People Analytics: How Social Sensing Technology Will Transform Business and What It Tells Us about the Future of Work* (Upper Saddle River, NJ: FT Press / Pearson Education, 2013).

19. David Talbot, "Facebook's Emotional Manipulation Study Is Just the Latest Effort to Prod Users," *Technology Review*, July 1, 2014, www.technologyreview.com/news/528706 /facebooks-emotional-manipulation-study-is-just-the-latest-effort-to-prod-users (accessed May 5, 2015); LinkedIn job advertisement for a data analyst for GSN (Game Show Network), www.linkedin.com/jobs2/view/10214638 (accessed May 10, 2015).

20. See, e.g., Jennifer R. Whitson and Claire Dormann, "Social Gaming for Change: Facebook Unleashed," *First Monday* 16, no. 10 (2011), www.firstmonday.org/ojs/index.php /fm/article/view/3578/3058 (accessed Aug. 24, 2015).

21. "Scaling Learning Analytics across Institutions of Higher Education," *Educause*, May 2013, http://net.educause.edu/ir/library/pdf/SEI1302.pdf (accessed May 5, 2015). Marist College is the institution where I currently work, but I am not involved with this project in any form.

22. Boyd and Crawford explain that the phenomenon of Big Data relies on three different dimensions that interact: *technology* (having the latest and greatest technical power to collect, analyze and contrast large data sets); *analysis* (being able to find patterns and make conclusions); and *mythology* (a belief that big data will provide new knowledge and new understanding into the mysteries of life). Boyd and Crawford, "Critical Questions for Big Data," 662; danah boyd and Kate Crawford, "Six Provocations for Big Data," presented at the Oxford Internet Institute Symposium on the Dynamics of the Internet and Society, Oxford, UK, Sept. 21, 2011.

23. Boyd and Crawford, "Critical Questions for Big Data."

24. "There remains a mistaken belief that qualitative researchers are in the business of interpreting stories and quantitative researchers are in the business of producing facts. In this way, Big Data risks reinscribing established divisions in the long running debates about scientific method and the legitimacy of social science and humanistic inquiry." Boyd and Crawford, "Critical Questions for Big Data," 667.

25. See, the "spurious correlations" at http://tylervigen.com (accessed May 5, 2015).

26. See, e.g., C. J. Pannucci and E. G. Wilkins, "Identifying and Avoiding Bias in Research," *Plastic Reconstructive Surgery* 126, no. 2 (2010): 619–25.

27. See boyd and Crawford, "Critical Questions for Big Data"; Marc Kosciejew, "The Perils of Big Data," *Feliciter* 59 (2013): 32–35; and Andrejevic, "Big Data Divide." In addition, samples can be too big and too small. If a sample is too big, it might lead to statistically significant outcomes even though differences are small or not relevant, leading to significance when there is none. David S. Fay and Ken Gerow, "A Biologist's Guide to Statistical Thinking and Analysis," *WormBook: The Online Review of* C. elegans *Biology*, July 9, 2013, www.wormbook.org/chapters/www_statisticalanalysis/statisticalanalysis.html#sec6-1 (accessed July 9, 2015).

28. Boyd and Crawford, "Critical Questions for Big Data"; Gitelman and Jackson, introduction. Laura Kurgan maintains, "No matter what, data has already been represented. There is no such thing as raw data. Data has always been translated from one form into another form . . . And it is always collected with a point of view, there are all kinds of biases."

Krugan, "Mapping and its Discontents," Dec. 17, 2013, www.youtube.com/watch?v =XIYvXLybvtY&feature=youtu.be&t=24m47s (accessed Sept. 1, 2015). "No such thing as raw data. Data is designed," quotes a Mar. 12, 2014, Facebook post linking the talk at www .facebook.com/SpatialInformationDesignLab/posts/10152368484805337 (accessed Sept. 4, 2015). For more on these questions, see the essays in Gitelman, *"Raw Data" Is an Oxymoron.*

29. Results of a Google search for "unleash your data," www.google.com/webhp?sourceid =chrome-instant&ion=1&espv=2&ie=UTF-8#q=unleash%20your%20data; Bernard Marr, "How to Tame the Big Data Beast," *Entrepreneur*, Feb. 6, 2015, www.entrepreneur.com/article /242387 (both accessed May 5, 2015).

30. Boyd and Crawford, "Critical Questions for Big Data," 671–73.

31. Andrejevic, "Big Data Divide."

32. Andrejevic, "Big Data Divide." See also "The Personal Information Project," www .cccs.uq.edu.au/personal-information-project (accessed July 9, 2015).

33. Kosciejew, "Perils of Big Data"; Andrejevic, "Big Data Divide."

34. Andrejevic, "Big Data Divide," 1685. See also "Personal Information Project."

35. Boyd and Crawford, "Critical Questions for Big Data."

36. Ibid., 672.

37. Boyd and Crawford note that some researchers at companies, such as Jimmy Lin, a professor spending his sabbatical at Twitter, argue that academics should not even be investigating social media data sets because industry's resources, tools, and analysis abilities are so much greater. Ibid., 674.

38. Andrejevic, "Big Data Divide," 1683; Lev Manovich, "Trending: The Promises and the Challenges of Big Social Data," 2011, http://manovich.net/content/04-projects/066 -trending-the-promises-and-the-challenges-of-big-social-data/64-article-2011.pdf (accessed May 5, 2015), 10. For more on the concept of the digital divide, see Mark Warschauer, *Technology and Social Inclusion* (Cambridge, MA: MIT Press, 2004).

39. Andrejevic, "Big Data Divide," 1676.

40. Boyd and Crawford, "Critical Questions for Big Data." Data scientist has been named one of the best jobs of 2015 and the sexiest job of the twenty-first century. Louis Columbus, "Glassdoor's 25 Best Jobs in America for 2015 Includes Data Scientists and Software Engineers," *Forbes*, Feb. 8, 2015, www.forbes.com/sites/louiscolumbus/2015/02/08/glassdoors -25-best-jobs-in-america-for-2015-includes-data-scientists-and-software-engineers; Thomas H. Davenport and D. J. Patil, "Data Scientist: The Sexiest Job of the 21st Century," *Harvard Business Review*, Oct. 2012, https://hbr.org/2012/10/data-scientist-the-sexiest-job-of -the-21st-century (both accessed May 5, 2015).

41. Andrejevic, "Big Data Divide," 1676. "The very notion of a panoptic sort is premised on a power imbalance between those positioned to make decisions that affect the life chances of individuals . . . and those subjected to the sorting process" (1677). See also Oscar H. Gandy, *The Panoptic Sort: A Political Economy of Personal Information* (Boulder, CO: Westview Press, 1993).

42. Andrejevic, "Big Data Divide," 1678.

43. Boyd and Crawford, "Critical Questions for Big Data," 665.

44. Ibid.

45. Ibid., 665, citing P. Du Gay and M. Pryke, *Cultural Economy: Cultural Analysis and Commercial Life* (London: Sage, 2012), 12–13.

46. Andrejevic, "Big Data Divide," 1681.

47. Geoffrey Bowker, "The Theory/Data Thing, Commentary," *International Journal of Communication* 8, no. 2043 (2014): 1795–99, 1797.

48. Christine Croft, "The Limits of Big Data" *SAIS Review of International Affairs* 34, no. 1 (2014): 118.

Chapter 9 · Knowledge

1. See R. C. Schank and R. Abelson, *Scripts, Plans, Goals, and Understanding* (Hillsdale, NJ: Erlbaum, 1977).

2. *The Restaurant Game* (Jeff Orkin and Deb Roy), http://alumni.media.mit.edu/~jorkin /restaurant (accessed Feb. 27, 2015), 5th para. Orkin and Roy sought to generate around 10,000 play-throughs. This goal was reached by Sept. 2010.

3. In *Sharkrunners* (area/code), real sharks were outfitted with tracking devices. Players could use the real-time information gathered by the sharks to play the game and earn money for different research funds. While *Sharkrunners* still uses human players, ostensibly, a game could be created with just nonhuman players.

4. Karin Knorr-Cetina, *Epistemic Cultures: How the Sciences Make Knowledge* (Cambridge, MA: Harvard University Press, 1999); Holly Korbey, "Debunking the Genius Myth," *Mind/Shift* (KQED), Aug. 30, 2013, http://ww2.kqed.org/mindshift/2013/08/30/debunking -the-genius-myth (accessed July 9, 2015).

5. Knorr-Cetina, *Epistemic Cultures*, 1; James Gee, *What Games Have to Teach Us About Learning and Literacy* (New York: Palgrave Macmillan, 2003); James Gee and Elisabeth Hayes, "Nurturing Affinity Spaces and Game-Based Learning," in *Games, Learning, and Society: Learning and Meaning in the Digital Age*, ed. Constance Steinkuehler, Kurt Squire, and Sasha Barab (New York: Cambridge University Press, 2012). On Samba Schools, see Seymour Papert, *Mindstorms: Children, Computers and Powerful Ideas*, 2nd ed. (New York: Basic Books, 1993).

6. Knorr-Cetina, *Epistemic Cultures*, 8. Knorr-Cetina looks not just at the way knowledge is constructed, but the "machinery" behind the creation of knowledge, including "different architectures of empirical approaches, specific constructions of the referent, particular ontologies of instruments, and different social machines . . . it brings out the *diversity* of epistemic cultures" (3).

7. See, e.g., Mark Andrejevic, "The Big Data Divide," *International Journal of Communication* 8 (2014): 1677, and danah boyd and Kate Crawford, "Critical Questions for Big Data: Provocations for a Cultural, Technological, and Scholarly Phenomenon," *Information, Communication, & Society* 15, no. 5 (2012): 662–79.

8. Mary Poovey, *A History of the Modern Fact: Problems of Knowledge in the Sciences of Wealth and Society* (Chicago: University of Chicago Press, 1998).

9. Boaz Miller, "When Is Consensus Knowledge Based? Distinguishing Shared Knowledge from Mere Agreement," *Synthese* 190 (2013): 1293–316, 1294.

10. Adam Frank, "Welcome to the Age of Denial," *New York Times*, Aug., 21, 2013, www .nytimes.com/2013/08/22/opinion/welcome-to-the-age-of-denial.html (accessed Feb. 25, 2015). See also Christine Aschwanden, "There's a Gap between What the Public Thinks and What Scientists Know," *FiveThirtyEight DataLab*, Jan. 29, 2014, http://fivethirtyeight.com /datalab/theres-a-gap-between-what-the-public-thinks-and-what-scientists-know (accessed Feb. 27, 2015).

11. Janis Dickinson and Rick Bonney, *Citizen Science: Public Participation in Environmental Research* (Ithaca, NY: Comstock, 2012), 3; Michael A. Nielsen, *Reinventing Discovery: The New Era of Networked Science* (Princeton, NJ: Princeton University Press, 2012), 6.

12. Nielsen, *Reinventing Discovery*, 3, 155.

13. Elizabeth Eisenstein, *The Printing Press as an Agent of Change* (Cambridge: Cambridge University Press, 1979), 518, citing Alexandra Koyré.

14. R. Louv, preface to Dickinson and Bonney, *Citizen Science*, x. The shift, Keeney explains, helped "free those at the pinnacle of the discipline from the time-consuming business of fieldwork that the then-prevalent natural-history focus of botany required." Elizabeth Keeney, *The Botanizers* (Chapel Hill: University of North Carolina Press, 1992), 23.

15. Dickinson and Bonney, *Citizen Science*, 8.

16. Ibid., 4. Although, we can also argue that these wicked problems always existed but we are now aware of them.

17. The "new kind of collaborative venture in data collection had been set in motion even before laboratory facilities were built or new observational instruments had been invented. The shift from script to print helps to explain why old theories were found wanting and new ones devised even before telescopes, microscopes, and scientific societies had appeared." Eisenstein, *Printing Press as an Agent of Change*, 520.

18. Nielsen, *Reinventing Discovery*, 183.

19. Louv, preface.

20. "We claim that the mere fact that participants have access to and control the experiential knowledge shapes and defines the nature of their relationships whether they are aware of it or not. They may refuse to participate, negotiate the level of their participation, and/or negotiate the emerging meanings of discourse developing through the data collection process, be it interview or observation." Adital Ben-Ari and Guy Enosh, "Power Relations and Reciprocity: Dialectics of Knowledge Construction," *Qualitative Health Research* 23 (2013): 423.

21. Ibid.

22. According to a Close the Door study and the *Guardian*, stores lose 91kg of carbon dioxide each week. See Will Nichols, "Shut That Door, Warns New Report from Cambridge University," *Guardian*, Nov. 25, 2010, www.theguardian.com/sustainable-business/close-the-door-cambridge-university, and "Facts & Research," www.closethedoor.org.uk/about-us/our-research (both accessed May 5, 2015).

23. Eric Hartman and Antoinette Hertel, "Clearer Thinkers, Better People? Unpacking Assumptions in Liberal Education," panel dialogue at Reinventing Liberal Education, Miami, FL, Nov. 22–23, 2013, www.nyu.edu/frn/publications/reinventing.liberal.education/Hartman.Hertel.html (accessed Feb. 27, 2015).

24. Boyd and Crawford, "Critical Questions for Big Data," 672.

25. On the historic role of the university, see, for example, Jürgen Habermas, *Toward a Rational Society: Student Protest, Science, and Politics* (Boston: Beacon Press, 1970).

26. Jessica Kleiman, "Why Getting a Liberal Arts College Education Is Not a Mistake," *Forbes*, Apr. 28, 2014, www.forbes.com/sites/work-in-progress/2014/04/28/why-getting-a-liberal-arts-college-education-is-not-a-mistake; John Tierney, "Career-Oriented Education vs. the Liberal Arts," *Atlantic*, Nov. 20, 2013, www.theatlantic.com/education/archive/2013/11/career-oriented-education-em-vs-em-the-liberal-arts/281640 (both accessed May 5, 2015).

27. Deborah Strumsky, Jose Lobo, and Joseph A. Tainter, "Complexity and the Productivity of Innovation," *Systems Research and Behavioral Science* 27, no. 5 (2010): 496–509.

28. John Ziker, "The Long, Lonely Job of Homo academicus," *Blue Review*, Mar. 31, 2014, https://thebluereview.org/faculty-time-allocation; Colleen Flaherty, "So Much to Do, So Little Time," *Inside Higher Ed*, Apr. 9, 2014, www.insidehighered.com/news/2014/04/09/research-shows-professors-work-long-hours-and-spend-much-day-meetings; "Contingent Appointments and the Academic Profession," www.aaup.org/report/contingent-appointments-and-academic-profession (all accessed May 5, 2015); boyd and Crawford, "Critical Questions for Big Data," 23.

29. Although the rise in MOOCs is not necessarily directly or only due to these concomitant changes.

30. Alan D. Sokal, "Transgressing the Boundaries: Towards a Transformative Hermeneutics of Quantum Gravity," www.physics.nyu.edu/faculty/sokal/transgress_v2/transgress_v2_singlefile.html; Richard Van Noorden, "Publishers Withdraw More Than 120 Gibberish Papers," *Nature*, Feb. 4, 2014, www.nature.com/news/publishers-withdraw-more-than-120-gibberish-papers-1.14763; John Bohannon, "Who's Afraid of Peer Review?" *Science*

342, no. 6154 (Oct. 4, 2013): 60–65, www.sciencemag.org/content/342/6154/60.full (all accessed May 5, 2015).

31. Matthew T. Marino and Michael T. Hayes, "Promoting Inclusive Education, Civic Scientific Literacy, and Global Citizenship with Videogames," *Cultural Studies of Science Education* 7 (2012): 945–54. They explain that "citizenship should entail more than simply understanding or engaging in science and technology knowledge and should involve an activist orientation suitable for participatory democracy" (947).

32. Yupanqui J. Muñoz and Charbel N. El-Hani, "The Student with a Thousand Faces: From the Ethics in Video Games to Becoming a Citizen," *Cultural Studies of Science Education* 7 (2012): 909–43, 914. See also Karen Schrier, "Designing and Using Games to Teach Ethics and Ethical Thinking," in *Learning, Education, and Games, Vol. 1: Curricular and Design Considerations*, ed. Karen Schrier (Pittsburgh, PA: ETC Press, 2014).

33. On good citizenship, see Eleanor Roosevelt, "Good Citizenship: The Purpose of Education," *Pictorial Review* 31 (Apr. 1930): 4, 94, 97, www.gwu.edu/~erpapers/documents/articles/goodcitizenship.cfm (accessed May 5, 2015), and Muñoz and El-Hani, "Student with a Thousand Faces." I define ethical thinking as a type of literacy or constellation of skills and affective and thought processes related to determining how to act ethically and how to think through ethical scenarios. The skills, concepts, and processes related to ethical thinking could include, for example, perspective taking, consideration of another's emotions, interpretation of evidence, or reflection on one's personal ethics. See more at Karen Schrier, "Designing and Using Games to Teach Ethics and Ethical Thinking," and Schrier, "Ethical Thinking and Sustainability in Role-Play Participants: A Preliminary Study," *Simulation & Gaming*, published online before print, Dec. 17, 2014, http://sag.sagepub.com/content/early/2014/11/20/1046878114556145.refs (accessed Aug. 19, 2015).

34. Muñoz and El-Hani, "Student with a Thousand Faces," 913; Christine M. Bachen, Pedro F. Hernández-Ramos, and Chad Raphael, "Simulating REAL LIVES: Promoting Global Empathy and Interest in Learning through Simulation Games," *Simulation & Gaming* 43, no. 4 (2012): 437–60.

35. Schrier, "Emotion, Empathy, and Ethical Thinking in *Fable III*."

36. Miguel Sicart, *The Ethics of Computer Games* (Cambridge, MA: MIT Press, 2009). On emergent aspects of play, see Katie Salen and Eric Zimmerman, eds., *The Game Design Reader: A Rules of Play Anthology* (Cambridge, MA: MIT Press, 2005), and Katie Salen and Eric Zimmerman, *Rules of Play* (Cambridge, MA: MIT Press, 2003). On values and play, see also Mary Flanagan and Helen Nissenbaum, *Values at Play in Digital Games* (Cambridge, MA: MIT Press, 2014).

37. Sicart, *Ethics of Computer Games*, 121.

38. Knorr-Cetina, *Epistemic Cultures*.

39. "Games can play a role in . . . understanding of science as a body of knowledge and a way of knowing." Muñoz and El-Hani, "Student with a Thousand Faces," 913.

40. Tippens and Jensen investigate the game *Citizen Science* and argue that it is impossible to fully simulate the complexity of overlapping ecologies in a game, and that instead it needs to be simplified somewhat, resulting in some aspects of the dynamism of reality being lost. This has been understudied, however. See more in Deborah J. Tippins and Lucas John Jensen, "Citizen Science in Digital Worlds: The Seduction of a Temporary Escape or a Lifelong Pursuit," *Cultural Studies of Science Education* 7 (2012): 851–56.

41. Matthew J. Gaydos and Kurt D. Squire, "Role Playing Games for Scientific Citizenship," *Cultural Studies of Science Education* 7 (2012): 821–44; Marino and Hayes, "Promoting Inclusive Education," 947.

42. "Video games have a power of simulation in which they provide fictional worlds within which the subjects can interact in a substantially deeper way than in the case of most

other media and, moreover, these worlds exhibit relationships with the realities lived by the players. These worlds created by video games are full of processes and contents of an ethical nature, which call into play the moral virtues of the player-subject during the gaming experience and in her relationship with peers in communities of players. Thus, video games can contribute to the education of players as citizens." Muñoz and El-Hani, "The Student with a Thousand Faces," 938.

43. Eisenstein, *Printing Press as an Agent of Change*.

44. Mia Consalvo, "Rule Sets, Cheating, and Magic Circles: Studying Games and Ethics," *International Review of Information Ethics* 4, no. 12 (2005): 9, 10, citing Huizinga; Salen and Zimmerman, *Rules of Play*; Stenros, "In Defence of a Magic Circle: The Social, Mental and Cultural Boundaries of Play," *DIGRA* 1, no. 2 (2014), http://todigra.org/index.php /todigra/article/view/10; Eric Zimmerman, "Jerked around by the Magic Circle," Feb. 7, 2012, www.gamasutra.com/view/feature/135063/jerked_around_by_the_magic_circle_.php (both accessed May 5, 2015); Jesper Juul, "The Magic Circle and the Puzzle Piece," in *Conference Proceedings of the Philosophy of Computer Games 2008* (Potsdam: Potsdam University Press, 2008).

45. Juul, "Magic Circle and the Puzzle Piece"; Consalvo, "Rule Sets, Cheating, and Magic Circles," 9, 10; Mia Consalvo, "There Is No Magic Circle," *Games and Culture* 4, no. 4 (2009): 408–17, http://remotedevice.net/ctcs-505/mia_2009.pdf (accessed Mar. 1, 2015).

46. This question was posed and discussed by Chris Garrett, Doug Maynard, and other members of the Hudson Valley Game Developers meet-up group on July 30, 2015.

47. Tanya Lewis, "Google's Artificial Intelligence Can Probably Beat You at Video Games," *LiveScience*, Feb. 25, 2015, www.livescience.com/49947-google-ai-plays-videogames .html (accessed May 5, 2015).

48. Harlan Ellison, "I Have No Mouth and I Cannot Scream" (1967), http://hermiene.net /short-stories/i_have_no_mouth.html (accessed Feb. 27, 2015), sec. 3, para. 6.

49. Ibid., sec. 5, para. 4.

50. Ibid., sec. 8, para. 3; Johann Huizinga, *Homo Ludens: A Study of the Play-Element in Culture* (Boston: Beacon Press, 1955), quoted in Casey O'Donnell, "Guest Editorial Preface, Special Issue on Meaningful Play, in Search of Hopeful Monsters: A Special Issue from International Conference on Meaningful Play," *International Journal of Gaming and Computer-Mediated Simulations* 7, no. 3 (2015): iv–vii, iv.

51. Jane McGonigal, *Reality Is Broken* (New York: Penguin, 2011); Ian Bogost, "Reality Is Alright: A Review of Jane McGonigal's Book *Reality is Broken*," http://bogost.com/blog /reality_is_broken (accessed May 5, 2015).